Lucia Panizzi

On a mathematical model for case hardening of steel

Lucia Panizzi

On a mathematical model for case hardening of steel

Modeling, analysis and numerical simulations

Südwestdeutscher Verlag für Hochschulschriften

Impressum/Imprint (nur für Deutschland/ only for Germany)
Bibliografische Information der Deutschen Nationalbibliothek: Die Deutsche Nationalbibliothek verzeichnet diese Publikation in der Deutschen Nationalbibliografie; detaillierte bibliografische Daten sind im Internet über http://dnb.d-nb.de abrufbar.

Alle in diesem Buch genannten Marken und Produktnamen unterliegen warenzeichen-, marken- oder patentrechtlichem Schutz bzw. sind Warenzeichen oder eingetragene Warenzeichen der jeweiligen Inhaber. Die Wiedergabe von Marken, Produktnamen, Gebrauchsnamen, Handelsnamen, Warenbezeichnungen u.s.w. in diesem Werk berechtigt auch ohne besondere Kennzeichnung nicht zu der Annahme, dass solche Namen im Sinne der Warenzeichen- und Markenschutzgesetzgebung als frei zu betrachten wären und daher von jedermann benutzt werden dürften.

Verlag: Südwestdeutscher Verlag für Hochschulschriften GmbH & Co. KG
Dudweiler Landstr. 99, 66123 Saarbrücken, Deutschland
Telefon +49 681 37 20 271-1, Telefax +49 681 37 20 271-0
Email: info@svh-verlag.de
Zugl.: Berlin, Technische Universitaet Berlin, Diss., 2010

Herstellung in Deutschland:
Schaltungsdienst Lange o.H.G., Berlin
Books on Demand GmbH, Norderstedt
Reha GmbH, Saarbrücken
Amazon Distribution GmbH, Leipzig
ISBN: 978-3-8381-2389-9

Imprint (only for USA, GB)
Bibliographic information published by the Deutsche Nationalbibliothek: The Deutsche Nationalbibliothek lists this publication in the Deutsche Nationalbibliografie; detailed bibliographic data are available in the Internet at http://dnb.d-nb.de.

Any brand names and product names mentioned in this book are subject to trademark, brand or patent protection and are trademarks or registered trademarks of their respective holders. The use of brand names, product names, common names, trade names, product descriptions etc. even without a particular marking in this works is in no way to be construed to mean that such names may be regarded as unrestricted in respect of trademark and brand protection legislation and could thus be used by anyone.

Publisher: Südwestdeutscher Verlag für Hochschulschriften GmbH & Co. KG
Dudweiler Landstr. 99, 66123 Saarbrücken, Germany
Phone +49 681 37 20 271-1, Fax +49 681 37 20 271-0
Email: info@svh-verlag.de

Printed in the U.S.A.
Printed in the U.K. by (see last page)
ISBN: 978-3-8381-2389-9

Copyright © 2011 by the author and Südwestdeutscher Verlag für Hochschulschriften GmbH & Co. KG and licensors
All rights reserved. Saarbrücken 2011

Contents

Introduction   iii

1 Physical background and modelling   1
    1.1 Phase transitions in iron-carbon equilibrium diagram   1
    1.2 Time-temperature-transformation diagrams   3
    1.3 Heat treatment of steel: case hardening   10
    1.4 Modelling of gas carburizing   11
        1.4.1 Carbon diffusion in austenite   11
        1.4.2 Phase transformations   14
        1.4.3 Heat transfer   15
        1.4.4 Complete model of gas carburizing   17

2 Analysis of the complete model   21
    2.1 Problem statement and strategy   21
    2.2 Assumptions and main results   22
    2.3 Proof of existence   24
    2.4 Proof of uniqueness   40

3 Analysis of a related quasilinear parabolic system   45
    3.1 Problem statement and strategy   45
    3.2 Main results   47
    3.3 Proof of existence   50

3.4	Proof of regularity	53
3.5	An anisotropic embedding theorem	66
3.6	Proof of continuous data dependence	72

4 Numerical results — 79

4.1	Introduction	79
4.2	Carburization	83
4.3	Quenching	88
4.4	Simulation of the complete process	93

5 Conclusions — 99

Bibliography — 102

Introduction

Outline

Despite the creation of numerous new functional materials, steel is still the basic material for the sustainable development of modern industrial society. Its applications are very diverse and widespread in all major branches of industry. Heat treatment of steel is a fundamental process which dates back thousands of years. Thanks to the tremendous flexibility of steel to various kinds of treatments, it is now possible to produce it with a large variety of desired properties.
The basic principle involved in heat treatment is the process of heating and cooling. In steel for example, hardness can be achieved by heating followed by rapid cooling. In general, two important properties of ferrous materials are contact fatigue strength and wear resistance, which depend mainly on the physical and chemical properties of a superficial layer. A special treatment acting on a relatively thin superficial layer of steel workpieces is called case hardening, because its aim is to harden just the workpiece case (i.e an encasing layer), letting the inner part softer. In nowadays high-technology industry the approach to case hardening often involves trial and error methods, based on previous experiences and empirical analysis. This kind of procedure requires costly and time-consuming experiments.

Due to its importance, there is a huge technical literature on the subject of heat treatment of steel, mostly of engineering type. On the other hand, the thermal processing of steel has also attracted the attention of mathematicians. The mathematical description of solid-solid phase transitions in steel started with the seminal works of Avrami [4] and Kolmogorov [30] in the thirties of last century. Since then the subject has been widely studied and, stimulated by the development of ever-faster computer hardware, numerous papers were published on the numerical simulation of the diffusion controlled phase transitions in steel. The first analytical investigation of phase transitions in steel, concerned with austenite-pearlite transition, is reported by A. Visintin [49]. Similar models have been studied in connection with polymerisation models [2]. The amount of works dedicated to the modelling of the phase transitions in steel has been increasing in the last decades, due to the numerous applications in industry. Studies about phase transitions in metallic alloys have been particularly undertaken at WIAS (*Weierstrass Institute for Applied Analysis and Stochastics*), where this thesis has been written, exploiting the large experience accumulated on the subject.

From the modelling point of view, heat treatments of steel include heat transfer and transition processes involving the many different crystalline species in steel. Occasionally, as in the specific case of carburizing processes, more equations are needed to describe the evolution of added chemicals.

The heat treatment that we want to investigate in the present work is the most widely used variant of case hardening, named gas carburizing. Carbon is indeed the key to the hardening of steel by the heating and quick cooling mechanism. If the used steel does not contain sufficient carbon to provide the required hardness, then its composition is altered only on the surface layer so that it can become hard during subsequent cooling. In carburizing, the workpiece is heated up to a certain temperature, then it is brought in contact with an environment of sufficient carbon potential to cause the carbon absorption at the surface and, by diffusion favoured by the high temperature, to create a carbon concentration gradient inside the workpiece. Then the workpiece is rapidly cooled down, so that the carbon diffusivity drops practically to zero and carbon atoms remain frozen within the iron lattice. This causes an atomic disorder and results in distortion of the lattice which manifests itself in the form of hardness and/or strength. Regarding this process, we do not know any reference to previously formulated models, though the literature dedicated to modelling phase transition in steel during heat treatments (like induction or flame hardening) is quite large (see the remarkable group of papers [19], [20], [21], [23]). Within the engineering literature, the process of gas carburizing is divided and investigated in two distinct parts: first carburization, then cooling. In the studies regarding the carburization stage, the focus is almost exclusively on the carbon evolution, described through a diffusion equation often with constant coefficients and disregarding of the coupling with temperature. This approach does not permit, for instance, to take into account further diffusion of carbon into the workpiece in a lower temperature range during the stage preceding quenching, which nevertheless is very important for the control of the carbon profile and the accuracy of the whole process. Among them, we quote for example [9], [12], [16], [44], [45] and [46]. A huge amount of studies regarding exclusively the quenching of steel in correlation with its hardening is available (see for instance [13], [25], [39] and the monograph [48]). To the best of our knowledge a complete model including all the possible effects has not yet been examined.

Objective

The aim of the present thesis is the development and analysis of a mathematical model for the process of gas carburizing, taking into account the basic characteristics under the actual conditions of the industrial process and relating the quality of the treatment with the technological parameters of the process. In view of the lack of mathematical studies concerning this process, we start with an analytical study, which focuses on the examination of the mathematical model and on the related mathematical questions. Then we present some numerical simulations to illustrate the applicability of our model to a concrete example.
The work is organised as follows.

Chapter 1 starts with some fundamental concepts regarding steel from the point of view of materials science and with the description of its most important metallurgical properties. We focus on a special type of case hardening process, named gas carburizing and come to the formulation of a mathematical model for this process, at the macroscopic level, in term of a set of differential equations, one for the diffusion of carbon, one for the thermal evolution, both coupled with the equations describing the evolution of the phase fractions, which are, in our case, two ordinary differential equations.

In Chapter 2 we analyse the complete model presented in the previous chapter. We examine the question of the existence of a weak solution of the corresponding initial-boundary value problem and consider the question of the uniqueness of the solution. As in the case of many evolutionary problems arising from the physics, the problem consists of a second order parabolic initial boundary value problems and requires the treatment of a quasilinear problem. In order to avoid further complications we have neglected all the mechanical effects (like e.g. shrinking accompanying cooling, stress analysis, etc.).

In order to deepen the analysis of some challenging questions which have arisen, we dedicate Chapter 3 to deepening the mathematical investigation. We concentrate therefore on a sub-problem strictly related to the one examined in Chapter 2. On the domain $Q_T = \Omega \times (0,T)$, with $\Omega \subset \mathbb{R}^N$ we consider the parabolic problem:

$$\begin{aligned}
\frac{\partial \theta}{\partial t} - \operatorname{div}(a(x,t)\nabla\theta) &= r(\theta,c) \\
\frac{\partial c}{\partial t} - \operatorname{div}\left(D(\theta,c)\nabla c\right) &= 0 \\
\frac{\partial \theta}{\partial \nu} + h(x,\theta,\theta_\Gamma(x,t)) &= 0 \\
-D(\theta,c)\frac{\partial c}{\partial \nu} &= b(x,t)
\end{aligned}$$

with prescribed initial conditions. They represent, respectively: heat transfer, carbon diffusion, heat exchange and carbon exchange. We address first the question of existence and L^∞-boundness of the solutions. We are particularly interested in the question concerning uniqueness and regularity of the solutions of the system above. The question of uniqueness is well known to be more involved than the one of existence. We are aware of the amount of (also very recent) results concerning this topic, which employ disparate techniques, enabling to treat very general cases in abstract settings, under very weak assumptions (see [15],[18] and references therein). Nevertheless, at present, a general theory covering all the possible cases is not available and every approach presents different advantages and disadvantages. A certain theory usually addresses particular issues and develops specific techniques to that aim. Some drawbacks are always present and render these approaches sometimes inapplicable to concrete cases. For instance, an undesired restriction on the dimension of the spatial domain often occurs.

After these considerations, we consider worthy to develop a technique which is self-contained and elementary as far as possible, in order to solve this problem. We have been able to work also in a three-dimensional setting, which is the relevant case for the applications, with very weak assumptions, which can realistically suit a number of application problems. From this point of view, the present study achieves some progresses in the theoretical analysis on quasilinear parabolic systems, pursuing a new technique - to our knowledge - to treat the question of uniqueness, regularity and continuous dependence on the data for a certain class of partial differential equations systems.

A fundamental tool employed in our approach to the problem are the anisotropic Sobolev spaces, which are found to be particularly adequate in the treatment of equations presenting different regularity in tangential and in normal directions, that is the case of our system with associated third-type boundary conditions. Besides this, we proved a refined version of the Gronwall lemma, that we believe to have some independent interest, to conclude the proof of uniqueness. In the chapter we came to prove a theorem that can be read as a generalisation of some embedding theorems for the classical Sobolev spaces, to anisotropic Sobolev spaces. Also this theorem and its corollaries should be of interest to the readership interested in parabolic systems.

In Chapter 4 we present some simulation work. The simulations, performed with realistic data parameters, have been implemented with a software based on finite elements and cover all the stages of the process and show a possible application of the model in a concrete situation. Finally we comment about possible further developments concerning both the mathematical aspects and applications.

A shorter and unified version of Chapters 1 and 2 has been accepted for publication in [22] and the content of Chapter 3 has been submitted in [32].

Chapter 1

Physical background and modelling

1.1 Phase transitions in iron-carbon equilibrium diagram

There is a wide choice of textbooks and monographs dealing with iron and steel in the context of physics and engineering. The description of the principles of heat treatment and hardening processes presented in this chapter is based on [3], [7] and on the monograph [48]. Iron is an allotropic metal, that is it can exist in more than one type of lattice structure depending upon temperature.

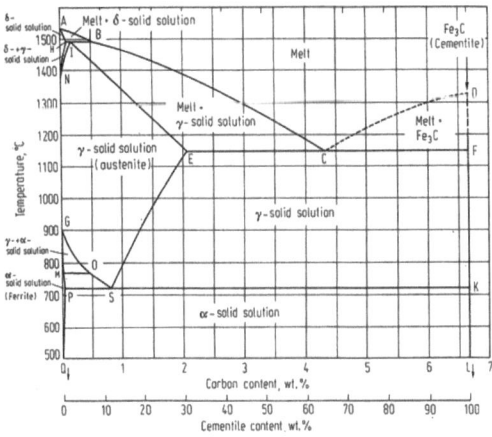

Figure 1.1: The iron-iron carbide equilibrium diagram, (from [48]).

The temperature at which the allotropic changes take place in iron is strongly influenced by alloying elements, the most important of which is carbon. The basis for a description of steel transformations is the diagram shown in Figure 1.1. It represents the metastable iron-iron carbide diagram, which is the part of interest, in practical applications, of the iron-carbon phase diagram. It is the part between pure iron and an interstitial compound of iron and carbon, called iron carbide or cementite, (denoted Fe_3C) containing 6.67 % carbon by weight. Therefore, this portion of the iron-carbon diagram is called iron-iron carbide equilibrium diagram. It is not a true equilibrium diagram, since equilibrium implies no change of phase with time, whereas the compound iron carbide decomposes into iron and carbon. Although iron-carbide is metastable, its decomposition takes a very long time at room temperature and under conditions of relatively slow temperature changes the system can be considered to cross quasi-equilibrium states represented in the iron-carbide diagram.

A *phase* is a portion of an alloy, physically, chemically, or crystallographically homogeneous throughout, which is separated from the rest of the alloy by distinct bounding surfaces. Phases that occur in iron-iron carbide alloys are austenite (γ phase), ferrite (α phase), cementite and graphite. They are shown in Figure 1.1. Besides these phases, there are many different structures in steel. On the basis of carbon content it is usual to divide the iron-carbon diagram into two parts. Those alloys containing less than 2 % carbon are known as *steel*, and those containing more than 2 % carbon are known as *cast irons*. The interesting region for many heat treatments is the one with carbon content less than 0.8%, the region of hypo-eutectoid steels. Steels with carbon content approximately 0.05-0.15% are called low-carbon steels. The microscopic structure of steel relies on the configuration of the iron lattice. Depending on temperature, two different lattice structures can occur: a body-centred-cubic (b.c.c.) and a face-centred cubic (f.c.c.) lattice. Above a certain temperature, steel is in the austenitic phase, the so-called γ-solid solution, an interstitial solid solution of carbon dissolved in f.c.c. iron. In this phase the solubility of carbon in iron is maximal. Below this temperature, the f.c.c. iron lattice is no longer stable. But before the lattice can change its configuration to form a b.c.c. structure, carbon atoms do diffuse, due to the fact that the solubility of carbon in b.c.c. iron is only about 1.4 % of the solubility in f.c.c. iron. The result is a lamellar aggregate of ferrite and cementite, called *pearlite*, where *ferrite* is a solid solution of carbon in b.c.c.. Figure 1.2 shows the two different lattice structures.

1.2. TIME-TEMPERATURE-TRANSFORMATION DIAGRAMS

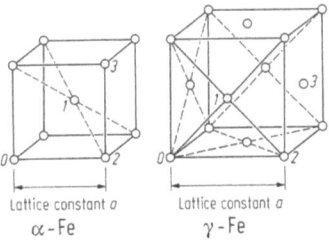

Figure 1.2: Crystal lattice structures of α-iron and γ-iron, (from [48]).

The transformation behaviour associated with various slow cooling rates is called diffusion-controlled because the carbon diffusion is dominating. The gamma-to-alpha transformation takes place by a process of nucleation and growth and is time-dependent. Increasing the cooling rate, if a critical cooling rate is overcome, the time for the carbon to diffuse out of solution is insufficient and the structure cannot become b.c.c. while the carbon is trapped in solution. The resultant structure, called *martensite*, is a solid solution of carbon trapped in a body-centred tetragonal structure.

The microstructure and distribution of different phases are of great importance, due to the fact that each of them possesses different hardness and mechanical properties. Martensite doesn't appear in diagram 1.1 since it is a metastable phase of steel, formed by a transformation of austenite below a certain temperature. It appears in the diagrams that will be shown in the next section.It is important for our aim to note that the difference in carbon solubility mentioned above makes it possible to produce defined microstructures and defined mechanical properties by technical heat treatment.

1.2 Time-temperature-transformation diagrams

From the iron-carbon diagram it is possible to derive some important facts regarding the properties of steel undergoing some heating or cooling. The kinetics of the phase transitions depicted in the previous section can be indeed described in an isothermal-transformation (IT) diagram, where temperature is plotted against time. Other names for the same curves are time-temperature-transformation (TTT) curves. These diagrams are constructed according to a standard procedure requiring heat treatment

CHAPTER 1. PHYSICAL BACKGROUND AND MODELLING

and metallographic experiments on samples. As a result of these experiments, two points can be plotted, namely, the time for the beginning and the time for the end of transformation. The entire experiment is repeated at different critical temperatures until sufficient points are determined to draw a curve showing the beginning of transformation. Portions of these lines are often shown as dashed lines to indicate a high degree of uncertainty.

Each different steel composition has its own TTT curve: an example is depicted in Figure 1.3. However, patterns are generally the same for all steels as far as shape of the curves is concerned. The most outstanding difference in the curves among different steels is the distance between the vertical axis and the nose of the S curve. This distance between the vertical axis and the nose of the S curve has a profound effect on how the steel must be cooled to form the hardened structure, martensite. The start of the transformation in the diagram is marked by the line at which 1% of the new microstructure is formed from the austenite. The curve for complete transformation is drawn for the point at which 1% of austenite is left. Above the critical temperature Ac_3, austenite is stable. The area on the left of the beginning of transformation consists of unstable austenite. The area on the right of the end of transformation line is the product to which austenite will transform at constant temperature. The area between the beginning and the end of the transformation consists of the microstructures: pearlite (P), ferrite (F), bainite (B) and martensite (M).

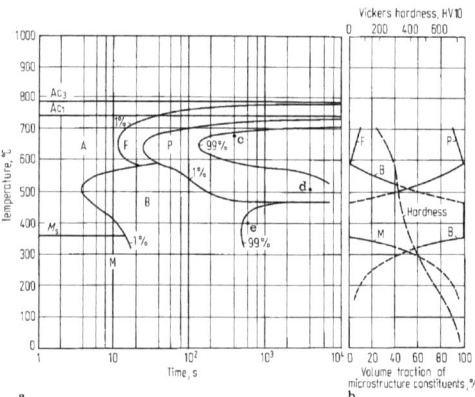

Figure 1.3: (a) Isothermal transformation diagram for the steel 41Cr4; (b) volume fraction of microstructures (solid lines) and Vickers hardness measured at room temperature (dotted line), (from [48]).

1.2. TIME-TEMPERATURE-TRANSFORMATION DIAGRAMS

The temperature at which the martensite transformation starts is denoted by M_s. It is almost impossible to determine experimentally with precision the temperature at which the martensite formation is finished, for this reason we can find very seldom time-temperature-transformation diagrams with the indication of M_f, but rather with M_{50} or M_{90}, denoting respectively the temperature at which the amount of formed martensite is 50% or 90%.

Phase transformations in steel, during different heat treatments, represent a wide field of investigation. For discussion concerning the material properties of steel in correlation with phase transformations and heat treatment, we refer to [3], [9] and [48].

As the austenite-pearlite transformation is a nucleation and growth process, it is governed by the nucleation rate (the amount of nuclei of the new phase formed per unit time and volume) and by the growth rate of the nuclei. Both of these factors depend on temperature, but in a different way: while the growth rate is high at high temperature and then decreases with decreasing temperature, the nucleation rate increases with decreasing temperature. This behaviour explains the 'nose shape' of the pearlite transformation curves.

Bainite is the product of a transformation which may take place, depending on the type of steel, in the temperature range between the pearlitic and martensitic phase. This additional transformation is shown very distinctly in the TTT diagrams of low-alloyed steels (see for instance Figure 1.3). Bainite is a microstructure made of carbide and martensite. It is generally difficult to recognize it with precision since the bainitic transformation presents similar features to the martensitic one and bainite presents a similar aspect to martensite (the german name for bainite 'Zwischenstufengefüge', means 'phase in between' and it is denoted by ZW).

In Figure 1.4, the microstructure of austenite is shown on the right. It is visible that the lower the temperature at which transformation takes place, the higher the hardness. It is also evident that all the structures from the top to the region where marteniste forms are time-dependent, but the formation of martensite is not time-dependent.

6 CHAPTER 1. PHYSICAL BACKGROUND AND MODELLING

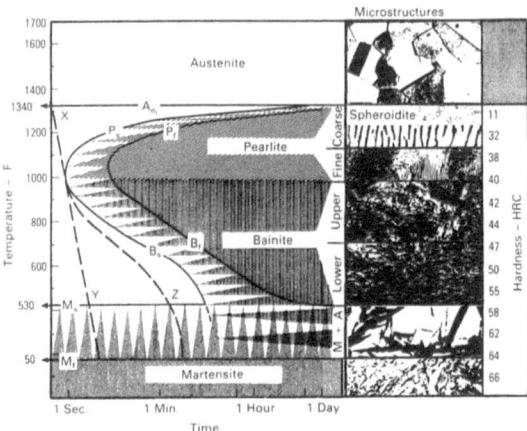

Figure 1.4: TTT diagram for an eutectoid carbon steel (0.77% C), (from [7]).

The austenite-martensite transformation is responsible for the outstanding position of steel, because it permits the extreme hardenability of this material. Concerning the *martensitic transformation*, which is important for the process that we are going to study, we list, below, its main properties.

- The martensitic transformation proceeds only during cooling and ceases if cooling is interrupted. If the steel is held at any temperature below the M_s, the transformation to martensite will stop and will not proceed again unless the temperature is dropped.

- Martensite can be considered as a transition structure from the unstable austenite phase to the final equilibrium condition of a mixture of ferrite and cementite. The most significant property of martensite is its great hardness; the distortion occurring in the lattice during cooling is its prime reason. Although martensite is harder than the austenite from which it forms, extreme hardness is possible only in steels containing sufficient carbon. The maximum hardness obtainable from steel in the martensitic phase is a function of carbon content only.

- The martensite transformation of a given alloy cannot be suppressed, nor can the M_s temperature be changed by changing the cooling rate. The temperature range of the formation of martensite is characteristic of a given alloy. Many formulas have been experimentally determined for the dependency of M_s upon

1.2. TIME-TEMPERATURE-TRANSFORMATION DIAGRAMS

composition of steels. According to [48], one is:

$$M_s(K) = 812 - 423(\%C) - 30.4(\%Mn) - 17.7(\%Ni) - 12.1(\%Cr) - 7.5(\%Mo)$$

The main dependency is clearly upon carbon concentration. The temperature M_s is lowered as the carbon content in the austenite increases. Theoretically, the austenite to martensite transformation is never complete, and small amounts of retained austenite will remain even at low temperatures. The transformation of the last traces of austenite becomes more and more difficult as the amount of austenite decreases. The influence of carbon on the M_s and M_f temperatures is shown in Figure 1.5.

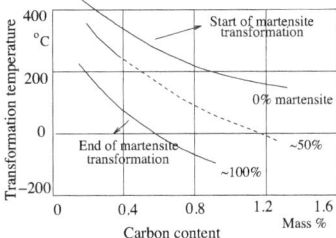

Figure 1.5: Influence of carbon on the martensite formation.

The IT diagram shows the time-temperature relationship for austenite transformation only as it occurs at constant temperature, but most heat treatments involve transformation on continuous cooling. Therefore, for the planning of heat treatments, it is important to predict the (nonisothermal) evolution of the phase fractions. This can be done by means of another type of diagram, the continuous-cooling-transformation (CCT) diagram, which describes the transformation of austenite during countinuous cooling. Figure 1.6 shows a CCT diagram for the steel whose TTT diagram has been previously shown (Fig. 1.3). As for the isothermal transformation, in general the start is defined as the temperature, at which 1% of the new microstructure has formed. The transformation is completed when only 1% of the original austenite is left. At each cooling curve these points are connected by smooth curves which represent the CCT diagram. The end of the formation of a microstructure is not clearly marked if another microstructure follows with decreasing temperature, as in the case of martensite after bainite. The end of the transformation in the martensite range is difficult to detect and therefore is not generally drawn.
Cooling curves are often characterised by the time taken for the specimen to cool from

800°C to 500°C or from Ac_3 to 500°C. CCT diagrams must be interpreted by following cooling curves. The start of the austenite transformation and the microstructures formed can be read from each cooling curve. Additional information on the volume fractions of the microstructures and the hardness of the specimen at room temperature are plotted as a function of cooling time in Figure 1.6 (b).

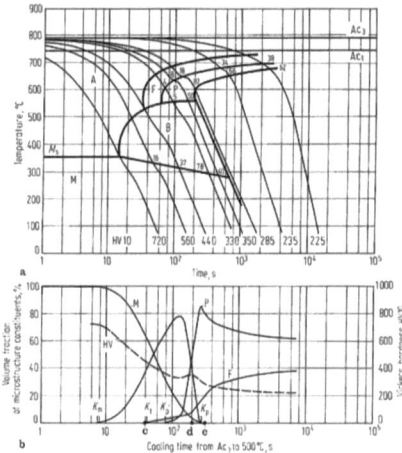

Figure 1.6: (a) Diagram for transformation during continuous cooling (CCT diagram) for the steel 41Cr4. The numbers on the cooling curves denote the volume fraction of microstructure constituents at room temperature; the numbers at the end of the cooling curves are the Vickers hardness at room temperature. (b) Volume fraction of microstructure constituents (solid lines) and Vickers hardness as function of cooling rate (dotted line) at room temperature, (from [48]).

Considering the IT diagram in relation to the location of the lines of the CCT diagram the 'nose' has been moved downward and to the right by continuous cooling.

The relation between the CCT diagram and the equilibrium diagram is plotted in Figure 1.7. With infinitely slow cooling or heating rate, the CCT diagram is identical to the equilibrium diagram for the composition of the steel. The equilibrium diagram is the limit to the transformation diagrams for infinite transformation time and infinitely slow cooling rate, respectively.

Figure 1.8 shows the relation between time, temperature and carbon content. The cooling curves are not depicted on this diagram for better readability. The line at 0.1 seconds shows the change of the M_s temperature depending on the carbon content.

1.2. TIME-TEMPERATURE-TRANSFORMATION DIAGRAMS

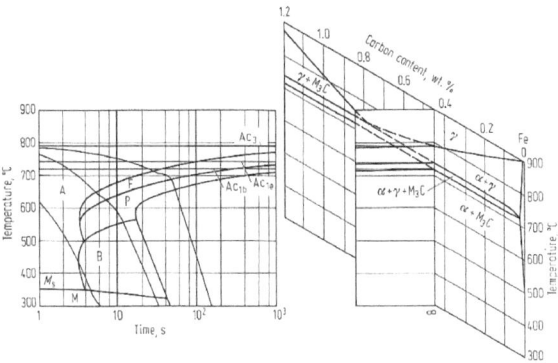

Figure 1.7: Equilibrium diagram of the system iron-carbon (right) as limitation of the CCT diagram of the steel 41Cr4 (left) with infinite low cooling rate, (from [48]).

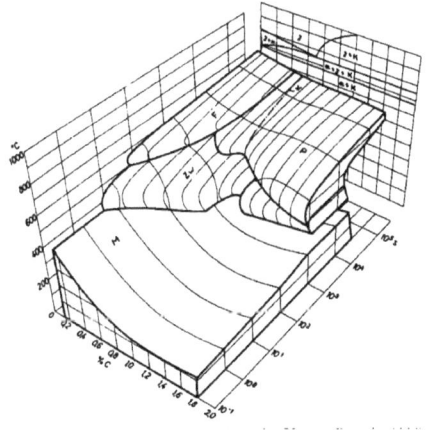

Figure 1.8: Three-dimensional presentation of the transformation characteristic of a 14NiCr14 steel, for continuous cooling, after austenization at 1023 K (Symbols: ZW bainite, M martensite, P pearlite, F ferrite), (from [48]).

Since we are interested in steels with carbon content varying from the core to the boundary, therefore diagrams as the one plotted in Figure 1.8 would also be needed in order to follow the transformation at all carbon levels, instead of a single CCT diagram for a steel with fixed carbon content. The cross sections for fixed carbon percentages give CCT diagrams like the one plotted in the left part of Figure 1.7. However, we note that such three-dimensional diagrams are very rare in literature and, as it will be shown in Chapter 4, other strategies are needed to determine the parameters entering in the equations describing the phase transitions.

1.3 Heat treatment of steel: case hardening

According to a definition given in [3], heat treatment of steel is a combination of heating and cooling operations, applied to a metal or alloy in the solid state in a way that will produce the desired properties.

All basic heat-treating processes for steel involve the transformation or decomposition of austenite. The nature and appearance of these transformation products determine the physical and mechanical properties of any given steel. All the different heat treatment processes consist of the following three stages, possibly cyclically repeated in different sequences:

1) heating of the material to some temperature in or above the critical range in order to form austenite;

2) holding the temperature for a certain period of time;

3) cooling, usually to room temperature.

The temperature and time for the various stages depend on the desired effect. During heating and cooling there exist temperature gradients between the outside and interior portion of the material, whose magnitude depends on the size and geometry of the workpiece.

All the heat treatments are associated with phase transformations in steel and indeed the phase transformations are the reason of the feasibility of the treatment itself. Generally, phase transformations may lead to mechanical deformation of the workpiece. Because of the mathematical complexity of the model that we are going to introduce, we will not include mechanical behaviour (e.g. thermo-elasticity, transformation induced plasticity) in our model. Therefore, distortion, in particular that related to the case hardening, can not be described by the model that we are going to propose.

This thesis is concerned with a particular kind of heat treatment, named *case hardening*, which is the production of parts that have hard, wear-resistant surfaces, but softer and/or tougher cores. It can be accomplished by two distinct methods. One approach is the use of a grade of steel that already contains sufficient carbon to provide the required surface after heat treatment (as flame or induction hardening). The surface

1.4. MODELLING OF GAS CARBURIZING 11

areas requiring the higher hardness are then selectively heated and quenched. The second method is the use of a steel that is not normally capable of being hardened to the desired degree, altering the composition of the surface layers by diffusion so that it either can be hardened during processing.

For practical purposes, case hardening of the second type can be broadly classified into four groups: carburizing, carbonitriding, nitriding and nitrocarburizing. Of these processes, carburizing is by far the most widely used.

1.4 Modelling of gas carburizing

1.4.1 Carbon diffusion in austenite

The primary object of carburizing is to provide a hard, wear-resistant surface with surface residual compressive stresses, that results in an improvement in the life of ferrous engineering components. In carburizing, austenitized ferrous metals are brought into contact with an environment of sufficient carbon potential to cause absorption of carbon at surface, creating, by diffusion, a concentration gradient between the surface and the interior of the metal.

Carburizing may be done in a gaseous environment (gas carburizing), a liquid salt bath (liquid carburizing), or with all the surfaces of the workpiece covered with a solid carbonaceous compound (pack carburizing). Regardless of the process, the objective of carburizing is to start with a relatively low-carbon steel (0.2% C) and to increase the carbon content in the surface layers, resulting in a high-carbon, hardenable steel on the outside with a gradually decreasing carbon content to the underlying layers. Process control to keep the desired carbon level is an important part of any carburizing process. Usually, although not always, approximately 0.8% is desired. Among the carburizing variants, the commercially most important is called *gas carburizing*, in which source of carbon is a carbon-rich furnace atmosphere: we will focus on this method. It is more effective, with deeper and higher carbon content cases obtained more rapidly and more adaptable for mass production than other carburizing processes.

During the process, the carbon-bearing atmosphere can be continuously replenished so that a high carbon potential can be maintained. A wide variety of furnaces are used for gas carburizing; selection of furnace type depends largely on workpiece shape and size, total production required and production flow.

It must be said that carburizing is a distinctly separate operation, taking place at constant temperature, high enough to have an austenitic microstructure, followed in most case by quenching, to form a martensitic microstructure, which produces the desired hardening effect. We will present a model for the complete process, including carburization and cooling.

The quality of the carburized samples is determined on the basis of the hardness and case depth required for a particular application and of compliance with given specifications and tolerances. Therefore it is necessary to understand the mechanism of carbon transfer and to accurately predict the carbon concentration profile and the case depth during heat treatment process.

The process of carburization can be divided in: transport of carbon to surface, reaction on the surface and diffusion of the carbon in the solid. The model we are going to present is a phenomenological description at the macroscopic level. Temperature and carbon concentration, in weight percent, are the physical variables playing the main role during carburization.

For an insight into aspects like atmosphere composition and kinetics of carbon transfer refer, for example, to [48].

The mechanisms of carbon transfer during carburizing is shown in Figure 1.9, where the most important physical parameters involved in the process are indicated: the mass transfer coefficient (β), defining carbon atoms flux (j) from the atmosphere through the steel surface, and the coefficient of carbon diffusion in steel (D) at austenitizing temperature.

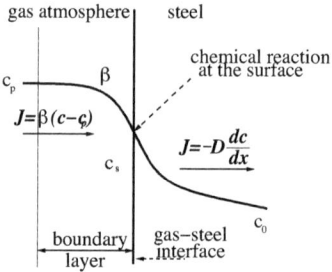

Figure 1.9: Schematic representation of carbon transport in carburizing.

The *mass transfer coefficient*, β, in the gas phase controls the rate of carbon uptake during the initial stage of carburizing. Several efforts to measure β have been made under variable temperature, atmosphere composition and carburizing potential. Ac-

1.4. MODELLING OF GAS CARBURIZING

cording to [16], [44], [45], its value ranges from $1 \cdot 10^{-9}$ to $2 \cdot 10^{-4}$ cm/s at temperatures between 800°C and 1000°C; nevertheless in many circumstances, it is not significant error assuming β constant.

The ability of the furnace atmosphere to supply carbon to the heated steel is called *carbon potential*, denoted with c_p (see Figure 1.9) and is defined as the percent carbon (by weight) in a piece of steel which has achieved thermodynamic equilibrium with the atmosphere. For example, if a part with 0.6% carbon neither takes carbon from the furnace atmosphere, nor gives up carbon to it, the atmosphere is said to have a carbon potential of 0.6%C. In such atmosphere, a part with 0.5%C would take carbon from the atmosphere (i.e. would be carburized) and a part with 0.7%C would be decarburized, (from [51]).

According to Fick's first law, the flow of substance is proportional to the concentration gradient

$$j = -D(\theta, c)\nabla c,$$

where the diffusivity $D = D(\theta, c)$ is a function of the concentration c and the temperature θ.

The unsteady process of diffusion of carbon in γ-iron (austenite) is described by the following parabolic equation

$$\frac{\partial c}{\partial t} - div(aD(\theta, c)\nabla c) = 0.$$

The factor a, denoting the austenite fraction, in front of the diffusion coefficient $D(\theta, c)$ reflects the fact that enrichment with carbon only takes place in the austenite phase. Now the boundary conditions must be assigned. Since the difference in carbon potential between the surface and the workpiece provides the driving force for carbon diffusion, we prescribe the following boundary condition:

$$-aD(\theta, c)\frac{\partial c}{\partial \nu} = \beta(\theta)(c - c_p)$$

where β, the mass transfer coefficient, controls the rate at which carbon is absorbed by the steel during carburizing and c_p is the carbon potential in the furnace. $\frac{\partial c}{\partial \nu}$ denotes the outward normal derivative. There are many formulas in literature for the carbon diffusivity in austenite, $D(\theta, c)$, according to the kinetic and thermodynamic behaviour of carbon in austenite. We refer to Chapter 4 for some comments about this coefficient.

1.4.2 Phase transformations

In this study, we take into account the phase transitions from austenite to pearlite and from austenite to martensite. In other words we assume that all the austenite present at the beginning of the transformation will be completely turned into pearlite and martensite, at the end. However, a more elaborate model accounting for all the phase transitions occurring during the heat treatment of steel can be found in [23], [50] and references therein.

Since the densities of the phases in steel differ only very little from each other (for fixed temperature), we do not distinguish between mass and volume fractions of the phases. Therefore, in the following, we will refer only to the 'phase fractions'.

As already mentioned before, in this study, we do not consider the heating stage and assume that cooling takes place from the high temperature phase austenite with phase fraction a to two different product phases, pearlite with fraction p and martensite with fraction m.

The model that we are going to present is a mathematical description of the phenomenologically observed phase transitions in steel. The basic assumption for our approach is that all the information about the evolution of phase transitions in a specific steel is contained in the isothermal and non-isothermal time-temperature-transformation diagrams. We propose the simplest phenomenological model able to describe our situation, that was developed, in the context of heat treatment problems, in a series of papers ([19], [20], [23]), to which we refer for an exhaustive discussion about modelling and for comparisons with other possible models. We consider the following equations to describe the evolution of p and m:

$$p_t = (1 - p - m)g_1(\theta, c) \qquad (1.4.1a)$$
$$m_t = [\overline{m}(\theta, c) - m]_+ g_2(\theta, c) \qquad (1.4.1b)$$
$$p(0) = 0 \qquad (1.4.1c)$$
$$m(0) = 0 \qquad (1.4.1d)$$

where c is the concentration of carbon. Here the bracket $[\ ]_+$ denotes the positive part function $[x]_+ = \max\{x, 0\}$ and the subscript t, as usual for PDEs, denotes the derivative with respect to time, which can be found in literature denoted with the dot (e.g. $\dot{p}$ instead of p_t).

While the growth rate of pearlite p_t is assumed to be proportional to the remaining austenite fraction, the rate of martensite growth m_t is zero if m exceeds either the non-perlitic fraction $1 - p$, or the threshold $\overline{m}$, depending on both temperature and

1.4. MODELLING OF GAS CARBURIZING

carbon concentration. The quantity $\overline{m}(\theta, c)$ represents the maximum attainable value of martensite fraction and is defined as

$$\overline{m}(\theta, c) = \min\{m_{KM}(\theta, c); 1-p\}.$$

The function m_{KM} describes the volume fraction of martensite according to the Koistinen-Marburger formula [29], i.e.,

$$m_{KM}(\theta, c) = \left(1 - e^{-c_{km}(c)(M_s(c)-\theta)}\right) H(M_s(c) - \theta),$$

where H is the heaviside function, c_{km} and M_s here are parameters depending on the carbon concentration and can be drawn from the respective TTT diagram. If during some stage of the heat treatment $\theta \geq M_s$, owing to the heaviside function, we have $m_t = 0$, whence no martensite is produced during this stage. c_{km} is a constant reflecting the transformation rate and varying with the steel composition. We note that, since $m_t = 0$, the irreversibility of the martensite transformation is incorporated in the model. The functions g_1 and g_2 are positive given functions that can be identified from the time-temperature-transformation diagrams described before. In Chapter 4 we will illustrate how this identification can be done for a specific steel. The final hardness is obtained by accumulating the contributions of the different phases formed along cooling.

1.4.3 Heat transfer

With the purpose of obtaining the temperature equation governing the whole process, we start considering the energy balance. For an incompressible, rigid body, it holds (see for instance [10]) the following balance of energy

$$\rho \frac{\partial e}{\partial t} + div\, \mathrm{q} = 0,$$

where e is the specific internal energy, q is the heat flux and ρ is the density. Considering the whole process, we assume that the specific internal energy has the form $e = \hat{e}(\theta, p, m)$, with a differentiable material function $\hat{e}$. Applying the Fourier's law we obtain

$$\mathrm{q} = -k(\theta)\nabla\theta,$$

where $k(\theta)$ is the heat conductivity of the material and we derive the following equation for the temperature

$$\rho\frac{\partial\hat{e}}{\partial\theta}\frac{\partial\theta}{\partial t} + \rho\frac{\partial\hat{e}}{\partial p}\frac{\partial p}{\partial t} + \rho\frac{\partial\hat{e}}{\partial m}\frac{\partial m}{\partial t} - div(k(\theta)\nabla\theta) = 0.$$

We denote the partial derivatives

$$\frac{\partial e}{\partial\theta} = \hat{\alpha}(\theta,p,m), \quad \frac{\partial e}{\partial p} = -\hat{L}_p(\theta,p,m), \quad \frac{\partial e}{\partial m} = -\hat{L}_m(\theta,p,m).$$

Concerning the material functions $\hat{\alpha}$, $\hat{L}_p$, $\hat{L}_m$, we follow the modelling of [23], where they only depend on θ: $\hat{\alpha}(\theta,p,m) := \alpha(\theta)$, $\hat{L}_p(\theta,p,m) := L_p(\theta)$, $\hat{L}_m(\theta,p,m) := L_m(\theta)$. Here $\alpha(\theta)$ represents the specific heat and $L_p(\theta)$ and $L_m(\theta)$ denote latent heats of the austenite-pearlite and the austenite-martensite phase changes, respectively. They are related to the enthalpy of the transformation and to the rate of transformation. Overall, we have obtained that the heat equation during the whole process can be written as

$$\rho\alpha(\theta)\frac{\partial\theta}{\partial t} - div\Big(k(\theta)\nabla\theta\Big) \;=\; \rho L_p(\theta)p_t + \rho L_m(\theta)m_t. \tag{1.4.2}$$

The heat source stems from the latent heats due to the phase transformations occurring during cooling. During the phase transitions from austenite to pearlite and from austenite to martensite, heat is released, therefore the functions $L_p(\theta)$ and $L_m(\theta)$ are always positive. Since the terms p_t, m_t, given through the formulas (1.4.1a),(1.4.1b) are positive as well, we can immediately see that the right-hand side of equation (1.4.2) is a positive quantity. We pass now to assigning the initial condition and the boundary condition, neglecting conduction and radiation effect:

$$-k(\theta,c)\frac{\partial\theta}{\partial\nu} \;=\; h(\theta - \theta_\Gamma)$$
$$\theta(x,0) \;=\; \theta_0(x).$$

Here θ_Γ is the external temperature, $\theta_0(x)$ is the temperature at the beginning of the process and h is the heat transfer coefficient. Since we do not take into account mechanical effects, we assume the density ρ to be constant.

In conclusion, the evolution of temperature during the process can be described by

1.4. MODELLING OF GAS CARBURIZING

the non-linear problem:

$$\rho\alpha(\theta)\frac{\partial \theta}{\partial t} - div\Big(k(\theta)\nabla\theta\Big) = \rho L_p(\theta)p_t + \rho L_m(\theta)m_t \qquad (1.4.3)$$

$$-k(\theta)\frac{\partial \theta}{\partial \nu} = h(\theta - \theta_\Gamma) \qquad (1.4.4)$$

$$\theta(x,0) = \theta_0(x). \qquad (1.4.5)$$

1.4.4 Complete model of gas carburizing

We collect now the equations from the previous sections. Let $\Omega \subset \mathbb{R}^3$ be an open bounded set with boundary $\partial\Omega$ and $Q_T := \Omega \times (0,T)$ the corresponding parabolic cylinder, where T represents the end time of the process. We obtain the following initial-boundary value problem.

$$\rho\alpha(\theta)\frac{\partial \theta}{\partial t} - div\Big(k(\theta)\nabla\theta\Big) = \rho L_p(\theta)p_t + \rho L_m(\theta)m_t \quad \text{in } Q_T \qquad (1.4.6)$$

$$\frac{\partial c}{\partial t} - div\Big((1-p-m)D(\theta,c)\nabla c\Big) = 0 \quad \text{in } Q_T \qquad (1.4.7)$$

$$p_t = (1-p-m)g_1(\theta,c) \quad \text{in } Q_T \qquad (1.4.8)$$

$$m_t = [\min\{\overline{m}(\theta,c); 1-p\} - m]_+ g_2(\theta,c) \quad \text{in } Q_T \qquad (1.4.9)$$

$$-k(\theta)\frac{\partial \theta}{\partial \nu} = h(\theta - \theta_\Gamma) \quad \text{on } \partial\Omega \times (0,T) \qquad (1.4.10)$$

$$-(1-p-m)D(\theta,c)\frac{\partial c}{\partial \nu} = \beta(\theta)(c - c_p) \quad \text{on } \partial\Omega \times (0,T) \qquad (1.4.11)$$

$$\theta(x,0) = \theta_0(x) \quad \text{in } \Omega \qquad (1.4.12)$$

$$c(x,0) = c_0(x) \quad \text{in } \Omega \qquad (1.4.13)$$

$$p(0) = 0 \quad \text{in } \Omega \qquad (1.4.14)$$

$$m(0) = 0 \quad \text{in } \Omega \qquad (1.4.15)$$

The central part of this thesis is the analytical investigation of the mathematical model presented above.

It permits to evaluate the effects of all the possible interactions between carbon diffusion, heat treatment and growth of phases in every stage of the process. Actually it would be capable to describe more general situations, for instance any other process composed of many cycles of heating and cooling, associated to the diffusion of a substance and to the evolution of temperature. Motivated by its relevance in industrial applications, we decided to concentrate on the analysis of the process of gas carburiz-

18   CHAPTER 1. PHYSICAL BACKGROUND AND MODELLING

ing.

In the previous sections, we were concerned with the physical phenomena occurring during case hardening. For clarity, now we summarise the process schedule and fix it in a schematic representation (Figure 1.10).

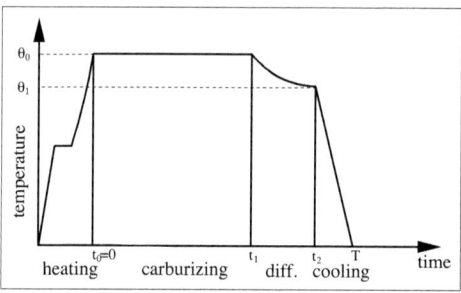

Figure 1.10: Schematic carburizing cycle.

We distinguish three characteristic times: the beginning of the carburization process t_0, the end of the carburization period t_1, the end of the diffusion period t_2 and the end of the cooling T.

First, the workpiece is heated up in a neutral atmosphere to a predetermined temperature θ_0 in the range from 1100 to 2150. During heating, steel is transformed to the austenite phase. Note that this period is not covered by the system (1.4.6)-(1.4.15).

Then a carbon rich gas such as propane, butane or methane is injected into the furnace. The workpiece is held at this temperature, to allow for the diffusion of carbon into the case. During this stage of the process, no phase transformation takes place, since the workpiece remains at the austenitizing temperature, well above the temperature where the formation of other phases can start. This fact is clearly observable in Figure 1.4. After the main carburizing stage, a slow diffusion period is sometimes allowed, through which the workpiece is brought to a lower specified temperature θ_1, whilst the carbon potential is lowered and the furnace is slowly cooled down. This diffusion step is used to remove the marked fluctuation in carbon concentration in the near-surface layer, which arises in the process of initial saturation. By allowing such a diffusion period, moreover, the surface carbon content can be reduced to any desired value. Again, in this further diffusive stage, the workpiece remains at a temperature high enough to prevent the formation of new phases, namely in the region where only the presence of austenite is possible.

After the diffusion period, the components are quenched to obtain the required hard-

1.4. MODELLING OF GAS CARBURIZING

ness and wear resistance on the surface. The duration of quenching is few minutes. Depending on the required depth of carburization, the total time of the process varies from one hour to several hours.

Of course for each different stage in which the whole process can be divided, as shown in Figure 1.10, we can consequently specify the coefficients involved in the boundary conditions:

- Stage 1: carburizing in a furnace, hence $\beta \geq 0$ and $h = 0$, since the carburizing takes place at constant temperature.

- Stage 2: diffusion period, with $\beta \geq 0$, c_p smaller than in Stage 1 and $h \geq 0$, serving as a linearized radiation law (a non linearized radiation law will be separately considered in Chapter 3).

- Stage 3: quenching with $\beta = 0$ and $h \geq 0$. During quenching the workpiece is taken out from the furnace and therewith from the carbon-reach atmosphere. This happens below the temperature necessary for the diffusion of carbon (previously named θ_1), therefore both further carburization and decarburization can be neglected. Hence we can assume β to be zero.

Chapter 2

Analysis of the complete model

2.1 Problem statement and strategy

This chapter is devoted to the analysis of the system (1.4.6)-(1.4.15), that we rewrite below for clarity:

$$\rho\alpha(\theta)\frac{\partial\theta}{\partial t} - \operatorname{div}\left(k(\theta)\nabla\theta\right) = \rho L_p(\theta)p_t + \rho L_m(\theta)m_t \quad \text{in } Q_T \qquad (2.1.1)$$

$$\frac{\partial c}{\partial t} - \operatorname{div}\left((1-p-m)D(\theta,c)\nabla c\right) = 0 \quad \text{in } Q_T \qquad (2.1.2)$$

$$p_t = (1-p-m)g_1(\theta,c) \quad \text{in } Q_T \qquad (2.1.3)$$

$$m_t = [\min\{\overline{m}(\theta,c); 1-p\} - m]_+ g_2(\theta,c) \quad \text{in } Q_T \qquad (2.1.4)$$

$$-k(\theta)\frac{\partial\theta}{\partial\nu} = h(\theta - \theta_\Gamma) \quad \text{on } \partial\Omega \times (0,T) \qquad (2.1.5)$$

$$-(1-p-m)D(\theta,c)\frac{\partial c}{\partial\nu} = \beta(c - c_p) \quad \text{on } \partial\Omega \times (0,T) \qquad (2.1.6)$$

$$\theta(x,0) = \theta_0(x) \quad \text{in } \Omega \qquad (2.1.7)$$

$$c(x,0) = c_0(x) \quad \text{in } \Omega \qquad (2.1.8)$$

$$p(0) = 0 \quad \text{in } \Omega \qquad (2.1.9)$$

$$m(0) = 0 \quad \text{in } \Omega \qquad (2.1.10)$$

The model accounts for the coupling of temperature, phase transitions and carbon diffusion. We will prove existence and uniqueness of a weak solution of system (2.1.1)-(2.1.10).

The proof of existence is carried out using a nested fixed point argument. We divide the proof in three steps. The first is a preliminary lemma concerning the ODE system

(2.1.3),(2.1.4),(2.1.9),(2.1.10) only, for θ and c prescribed. It is the case of a ODE system with discontinuous right-hand side, where the derivatives are defined almost everywhere. We will show that the conditions of the Charathéodory existence theorem are satisfied and, afterwards, that the solution is unique.

The second step is the coupling of the ODE system with the temperature equation, which gives a solution p, m, θ depending on c. The third and last step is the further coupling with the equation for c. To prove existence of a solution for the whole system, we proceed applying twice the Schauder fixed point theorem. To obtain uniqueness we must consequently develop an *ad hoc* technique.

2.2 Assumptions and main results

During this chapter we assume that $\Omega \subset \mathbb{R}^3$ and $\partial\Omega$ is its C^2-boundary and we denote $Q_T := \Omega \times (0,T)$ the corresponding space-time cylinder. During this and the following chapter we make use of the following notations:

- $W^{1,\infty}(0,T;L^\infty(\Omega)) = \{\, v \in L^\infty(0,T;L^\infty(\Omega)) \,:\, v_t \in L^\infty(0,T;L^\infty(\Omega)) \,\}$.

- $W_p^{r,s}(Q_T) = L^p(0,T;W_p^r(\Omega)) \cap W_p^s(0,T;L^p(\Omega))$.

- We denote by V the space $W_2^1(\Omega)$ and by V^* the space $(W_2^1(\Omega))^*$.
  $W(0,T) = \{\, v \in L^2(0,T;V) \,:\, v_t \in L^2(0,T;V^*) \,\}$, endowed with the norm
  $$\|v\|_{W(0,T)} = \left(\int_0^T (\|v(t)\|_V^2 + \|v'(t)\|_{V^*}^2) dt \right)^{\frac{1}{2}}.$$

The derivative of $u(x,t)$ with respect to t will be denoted, depending on the circumstances, with $\frac{\partial u}{\partial t}$ or with u_t.

Throughout the chapter we will make use of the following assumptions:

(A1) ρ and k are positive constants.

(A2) $\alpha \in C(\mathbb{R})$ and there exist positive constants α_0, α_1 such that $\alpha_0 \leq \alpha(\cdot) \leq \alpha_1$. $L_p, L_m \in L^\infty(\mathbb{R})$ and they are Lipschitz-continuous.

(A3) θ_Γ is a positive constant. $h \in W_2^2(\partial\Omega)$ with $h(x) \geq 0$ a.e. in $\partial\Omega$. $\theta_0 \in W_2^2(\Omega)$, $\theta_\Gamma \geq \theta_0$.

2.2. ASSUMPTIONS AND MAIN RESULTS

(A4) g_1, g_2 are Lipschitz-continuous in both arguments and there exist positive constants γ_1, γ_2 such that $0 \leq g_1(\theta, c) \leq \gamma_1$, $0 \leq g_2(\theta, c) \leq \gamma_2$, $\forall \theta, c \in \mathbb{R}$.

(A5) $\overline{m}$ is Lipschitz-continuous satisfying $\overline{m}(\theta, c) \in [0, 1]$ for every $\theta, c \in \mathbb{R}$.

(A6) $D(\theta, c)$ is Lipschitz-continuous in both arguments and there are positive costants γ_3, γ_4 such that $\gamma_3 \leq D(\theta, c) \leq \gamma_4$, $\forall \theta, c \in \mathbb{R}$.

(A7) c_p is a positive constant. $\beta \in L^\infty(\partial\Omega)$ with $\beta \geq 0$ a.e. in $\partial\Omega$. $c_0 \in L^2(\Omega)$.

We are going to prove that, under the hypothesis above, the considered problem has a weak solution.

Theorem 2.2.1 (Existence of a weak solution). *Assume (A1)-(A7), then there exists a weak solution (θ, c, p, m) to problem (2.1.1)-(2.1.10) such that $\theta \in W_2^{2,1}(Q_T)$, $c \in W(0, T)$, $p, m \in W^{1,\infty}(0, T; L^\infty(\Omega))$.*

The following proposition is also important from the physical point of view, since it provides a uniform bound on the temperature and on the carbon concentration, as we can expect from the process observation.

Proposition 2.2.2. *Assume (A1)-(A7). Moreover assume that $\theta_0 \in L^\infty(\Omega)$ and there exist positive constants C_1, C_2 such that $C_1 \leq c_p \leq C_2$ in $\partial\Omega \times (0, T)$ and $C_1 \leq c_0 \leq C_2$ a.e. in Ω. Then*

$$\theta_\Gamma \leq \theta \leq \theta_0 \quad \text{a.e in } Q_T$$

$$C_1 \leq c \leq C_2 \quad \text{a.e. in } Q_T.$$

Theorem 2.2.3 (Uniqueness). *Suppose that (A1)-(A7) are satisfied. Assume moreover that α is constant, $D = D(\theta)$, $h, \beta \in W_5^1(\partial\Omega), \theta_0, c_0 \in W_5^2(\Omega)$. Then the solution to (2.1.1)-(2.1.10) is unique.*

Remark 1. (a) Concerning Theorem 2.2.1: the regularity assumptions on the boundary and initial values could be weakened. To avoid unnecessary technicalities we assumed θ_Γ and c_p to be constants, but they could be in fact functions of space and time, according, for instance, to [37].

(b) Concerning Proposition 2.2.2: also in this case some weaker assumptions might be sufficient. Regarding θ_Γ and c_p, it is indeed sufficient their boundedness, but we preferred to keep the setting used for the existence theorem.

2.3 Proof of existence

As already mentioned, the proof is carried out using a nested fixed point argument. The first step is a preliminary lemma concerning the ODE system (2.1.3),(2.1.4) only, for θ and c prescribed. The second step is the coupling of the ODE system and the temperature equation, which gives a solution p, m, θ depending on c and the third is the further coupling with the equation for c. We begin with considering the initial value problem

$$z_t = f(z, \theta, c) \quad \text{in } Q_T \qquad (2.3.1a)$$
$$z(0) = 0 \quad \text{in } \Omega \qquad (2.3.1b)$$

where $z = (p, m)^T$ and $f = (f_1, f_2)^T$ denotes the right-hand side of (2.1.3),(2.1.4). The system of ODE's is meant as ODE with value in a Banach space, therefore, writing $z(x,t)$ with an abuse of notation, we mean $z(t)(x)$.

Lemma 2.3.1. *Under the assumptions (A4),(A5) the following statements are valid:*

(a) *For every $\theta, c \in L^2(Q_T)$ problem (2.3.1) has a unique solution z such that $p \geq 0$, $m \geq 0$ and*

$$|||z|||_{W^{1,\infty}(0,T;L^\infty(\Omega))} \leq M$$

for a constant M independent of θ and c. Moreover, there exists a constant c_T such that

$$0 \leq p(x,t) + m(x,t) \leq c_T < 1 \quad \text{for a.e. } (x,t) \text{ in } Q_T.$$

(b) *There are constants $M_1, M_2 > 0$ such that for every $\theta_1, \theta_2, c_1, c_2 \in L^p(Q_T)$, for almost all $t \in (0,T)$ and all $p \geq 2$ we have*

$$|||z_1(t) - z_2(t)|||_{W^{1,p}(\Omega)}^p \leq M_1 \int_0^t \|\theta_1 - \theta_2\|_{L^p(\Omega)}^p ds + M_2 \int_0^t \|c_1 - c_2\|_{L^p(\Omega)}^p ds \quad (2.3.2)$$

where p_i, m_i is the solution corresponding to (θ_i, c_i), and $|\cdot|$ is the Euclidean

2.3. PROOF OF EXISTENCE

norm in $\mathbb{R}^2$.

We will derive global existence on the interval (0,T) making use of the following classical result, which can be found for instance in [11, Theorem 1.1, Chap. 2]

Theorem 2.3.1 (Theorem of Carathéodory). *We consider the initial value problem:*

$$\begin{cases} u'(t) = F(u(t), t) & \text{on } J \\ u(t_0) = u_0 \end{cases} \quad (P)$$

where $J \subset \mathbb{R}$ with $t_0 \in J$ and the map F is of the form $F : K \times J \to \mathbb{R}^N$, where K is a compact set in $\mathbb{R}^N$. We assume:
(i) $F : K \times J \to \mathbb{R}^N$ satisfies the Carathéodory condition, i.e.

- $t \to F_i(u, t)$ *is measurable on J for each $u \in K$*

- $u \to F_i(u, t)$ *is continuous on K for almost all $t \in J$*

(ii) There exists a Lebesgue-integrable function $M : J \to \mathbb{R}$ such that $|F_i(t, u)| < M(t)$ for all $(t, u) \in J \times K$ and all i.

Then there exists a solution u of problem (P) defined on a neighbourhood U of t_0, 'in the extended sense', which means that u is absolutely continuous, $u(\cdot)$ satisfies (P) almost everywhere on U and satisfies the initial condition.

Now we can start the proof of Lemma 2.3.1.

Proof of Lemma 2.3.1 In order to prove (a) it is convenient to rewrite problem (2.3.1) as:

$$z_t = F(z, t) \quad \text{in } (0, T) \tag{2.3.3a}$$
$$z(0) = 0 \tag{2.3.3b}$$

with $F(z, \cdot) = f(z, \theta(\cdot), c(\cdot))$.

First of all we are going to show that the hypothesis $(i), (ii)$ of the Carathéodory's Theorem are satisfied, with $K = [0, 1] \times [0, 1]$ and $J = (0, T)$:

(i)

- $t \to F(z, t)$ is measurable on $(0, T)$ for each $z \in [0, 1] \times [0, 1]$;

- $z \to F(z, t)$ is continuous on $[0, 1] \times [0, 1]$ for almost all $t \in (0, T)$.

These conditions follow from the definition of F as a consequence of the measurability of θ and c on $(0,T)$ and of the fact that $g_1(\theta,c), g_2(\theta,c)$ are Lipschitz continuous in both variables.

(ii) Using assumption (A4),(A5) we have

$$|F_1(z,t)| \leq |1-p-m|g_1(\theta,c) \leq \gamma_1 \quad \text{on} \quad [0,1] \times [0,1] \times (0,T),$$

$$|F_2(z,t)| \leq |\overline{m}-m|g_2(\theta,c) \leq \gamma_2 \quad \text{on} \quad [0,1] \times [0,1] \times (0,T).$$

According to Carathéodory Theorem, system (2.3.3) has a solution in the extended sense on some time interval $[0,T_+)$, for some $T_+ > 0$.
Next we are going to show that the solution is globally unique. To this end it is sufficient to prove that there holds

$$|F(z_1,t) - F(z_2,t)| \leq L|z_1 - z_2| \quad \forall (z_1,t),(z_2,t) \in [0,1] \times [0,1] \times (0,T). \quad (2.3.4)$$

Indeed, according to the definition of F:

$$|F(z_1,t) - F(z_2,t)|^2 = |(1-p_1-m_1)g_1(t) - (1-p_2-m_2)g_1(t)|$$
$$+ \left|[\min\{\overline{m}(t);1-p_1\} - m_1]_+ g_2(t) - [\min\{\overline{m}(t);1-p_2\} - m_2]_+ g_2(t)\right|.$$

Thanks to the boundedness of g_1 and g_2, we obtain
$|(1-p_1-m_1)g_1(t) - (1-p_2-m_2)g_1(t)| \leq \gamma_1(|p_1-p_2| + |m_2-m_1|)$
and

$$|[\min\{\overline{m}(t);1-p_1\} - m_1]_+ g_2(t) - [\min\{\overline{m}(t);1-p_2\} - m_2]_+ g_2(t)|$$
$$\leq \gamma_2|\min\{\overline{m}(t);1-p_1\} - m_1 - \min\{\overline{m}(t);1-p_2\} + m_2|.$$

We shall now distinguish some cases.
If either $\min\{\overline{m}(t);1-p_i\} = 1-p_i$ or $\min\{\overline{m}(t);1-p_i\} = \overline{m}(t)$, for $i=1,2$, (2.3.4) immediately follows.
If $\min\{\overline{m}(t);1-p_1\} = 1-p_1$ and $\min\{\overline{m}(t);1-p_2\} = \overline{m}(t)$ (the same holds for inverted indices), we have $p_2 < 1 - \bar{m} < p_1$, therefore

$$\gamma_2|\min\{\overline{m}(t);1-p_1\} - m_1 - \min\{\overline{m}(t);1-p_2\} + m_2|$$
$$\leq \gamma_2(|m_1 - m_2| + |1 - p_1 - \overline{m}(t)|) \leq \gamma_2\Big(|m_1 - m_2| + |p_1 - p_2|\Big). \quad (2.3.5)$$

2.3. PROOF OF EXISTENCE

Thus, there exists a positive constant L such that

$$|F(z_1,t) - F(z_2,t)| \leq L|z_1 - z_2| \quad \forall (z_1,t), (z_2,t) \in [0,1] \times [0,1] \times (0,T).$$

Hence we have proved uniqueness of z on $(0,T)$.

To ensure global existence of the solution we need more an a priori estimate on an arbitrary subinterval of (0,T) (see for instance [52], pp. 799-800). To this aim we introduce the variable $Z = p+m$ and define T_ϵ as the maximal time such that $Z < 1-\epsilon$ on $(0, T_\epsilon)$.

We will show that for any $T > 0$ there exists an ϵ such that $0 \leq Z \leq 1-\epsilon$ in $(0,T)$. By this means, we will have also shown the last statement of part (a) of the lemma. This will be done by means of a classical comparison criterium for ODEs (see for instance [28, Prop. 3.1, Chap. 1]). Z satisfies, on $[0, T_\epsilon)$:

$$\dot{Z}(t) = (1 - Z(t))g_1(t) + [\min\{\overline{m}(t); 1 - p(t)\} - m(t)]_- g_2(t)$$
$$\leq g(t, Z(t)) := (1 - Z(t))(g_1(t) + g_2(t)),$$
$$Z(0) = 0.$$

Now, if we consider on $[0, T]$ the auxiliary problem:

$$\dot{V}(t) = (1 - V(t))(g_1(t) + g_2(t)) = g(t, V(t))$$
$$V(0) = 0$$

the solution is given by

$$V(t) = 1 - e^{-\int_0^t (g_1+g_2)(s)ds} \quad \forall t \in [0,T]$$

and we immediately have that there exists a constant $c_T > 0$ such that:

$$0 \leq V(t) \leq c_T < 1 \quad \text{on} \quad [0,T].$$

Notice that $g(t, V(t)) = (1 - V(t))(g_1(t) + g_2(t))$ is Lipschitz continuous on $[0, T]$ with respect to V. Thus, choosing $\epsilon = 1 - c_T$, we have

$$Z(t) \leq V(t) \leq 1 - \epsilon \quad \text{on} \quad [0,T].$$

Since T_ϵ was chosen maximally such that $Z(t) \leq 1 - \epsilon$ on $(0, T_\epsilon)$, it follows that

$T_\epsilon \geq T$.

Now, since $Z < 1 - \epsilon$, we have, by definition, that p_t and m_t are positive with $p(0) = m(0) = 0$. Therefore $p \geq 0$ and $m \geq 0$ on $(0, T)$. It follows that

$$0 \leq p + m \leq c_T < 1 \text{ on } (0, T).$$

Thus part (a) of Lemma 2.3.1 is concluded.

(b) Let us consider again the equation $z_t = f(z, \theta, c)$. Let z_i be the solution to the system (2.3.1), corresponding to θ_i, c_i, $i = 1, 2$. Denoting $z = z_1 - z_2$, subtracting the equations and taking the scalar product with the function $|z|^{p-2}z$, we obtain:

$$\frac{1}{p}\int_\Omega |z(t)|^p dx = \int_0^t \int_\Omega \Big(f(z_1, \theta_1, c_1) - f(z_1, \theta_2, c_2)\Big) \cdot z|z|^{p-2} dx ds. \qquad (2.3.6)$$

Invoking (A4), f is Lipschitz-continuous in all variables, thus, proceeding from (2.3.6), the conclusion follows through standard application of Young's inequality and Gronwall lemma. The proof of Lemma 2.3.1 is thus completed. ∎

Next, we define

$$B(\theta, c) := \rho L_p(\theta) p_t + \rho L_m(\theta) m_t, \qquad (2.3.7)$$

where (p, m) depends on θ, c as characterized by the previous lemma.

Lemma 2.3.2. *Suppose that (A2), (A4), (A5) hold. Then the operator B defined by (2.3.7) has the following properties*

(a) *There exists a constant $\bar{B}$ independent of θ, c such that, for all $\theta \in L^2(Q_T)$, $c \in L^2(Q_T)$ there holds*

$$\|B(\theta, c)\|_{L^\infty(Q_T)} \leq \bar{B}.$$

(b) *Given $c \in L^2(Q_T)$, let $\theta_k \subset L^2(Q_T)$ be any sequence converging strongly in $L^2(Q_T)$ to $\theta \in L^2(Q_T)$. Then for every $p \in [1, \infty)$, we have*

$$B(\theta_k, c) \to B(\theta, c) \quad \text{strongly in } L^p(Q_T). \qquad (2.3.8)$$

(c) *There are constants $K_1, K_2 > 0$ such that for all $\theta_1, \theta_2, c_1, c_2 \in L^2(Q_T)$ and for*

2.3. PROOF OF EXISTENCE

almost all $x \in \Omega$ and every $t \in (0,T)$

$$\int_0^t \left|B(\theta_1(x,s), c_1(x,s)) - B(\theta_2(x,s), c_2(x,s))\right|^2 ds$$

$$\leq K_1 \int_0^t |\theta_1(x,s) - \theta_2(x,s)|^2 ds + K_2 \int_0^t |c_1(x,s) - c_2(x,s)|^2 ds.$$

Proof of Lemma 2.3.2

(a) follows directly from assumptions (A2),(A4),(A5) and Lemma 2.3.1 (a).

(b) Emphasizing the dependence of p and m from the c and θ, as possible according to the previous lemma, we have

$$\frac{\partial p_{\theta,c}}{\partial t} = (1 - p - m) g_1(\theta, c)$$
$$\frac{\partial m_{\theta,c}}{\partial t} = [\min\{\overline{m}(\theta, c); 1 - p\} - m]_+ g_2(\theta, c).$$

Let $x \in \Omega \setminus N$, with $N \subset \Omega$ of zero measure and consider $z = (p,m)^T$. By Lemma 2.3.1 (a), $\| z_{\theta_k} \|_{W^{1,\infty}(0,T;L^\infty(\Omega))} \leq M \ \forall k$, thus $\| z_{\theta_k} \|_{W^{1,p}(0,T;L^\infty(\Omega))} \leq M \ \forall k, \ \forall p < \infty$. Thus, there exists a subsequence, $\{\theta_{k'}\}$, and some $\hat{z}$ such that

$$z_{\theta_{k'}}(x, \cdot) \to \hat{z}(x, \cdot) \quad weakly - star \quad in \quad W^{1,\infty}(0,T), \quad \text{for a.e. } x \in \Omega.$$

Thus, we have

$$\frac{\partial z_{\theta_{k'}}}{\partial t}(x, \cdot) \to \frac{\partial \hat{z}}{\partial t}(x, \cdot) \quad weakly \quad in \ L^p(0,T) \ \forall p < \infty, \text{ for a.e. } x \in \Omega.$$
$$z_{\theta_{k'}}(x, \cdot) \to \hat{z}(x, \cdot) \quad strongly \quad in \quad C[0,T], \quad \text{for a.e. } x \in \Omega.$$

Since the solution to system (2.3.1) is unique, we have $\hat{z}(x, \cdot) = z_\theta(x, \cdot)$ and the convergence holds for the whole sequence, hence we can conclude that $z_{\theta_k}(x,t) \to z_\theta(x,t)$ pointwise in Q_T. Since $\theta_k \to \theta$ strongly in $L^2(Q_T)$, using assumption (A4), possibly extracting a subsequence, we have

$$\rho L_p(\theta_{k'}) \frac{\partial p_{\theta_{k'}}}{\partial t} + \rho L_m(\theta_{k'}) \frac{\partial m_{\theta_{k'}}}{\partial t} \to \rho L_p(\theta) \frac{\partial p_\theta}{\partial t} + \rho L_m(\theta) \frac{\partial m_\theta}{\partial t} \quad \text{a.e in } Q_T. \quad (2.3.9)$$

But, applying Lebesgue theorem, we get

$$B(\theta_{k'}, c) \to B(\theta, c) \quad \text{strongly in } L^p(Q_T).$$

Since the limit does not depend on the extracted subsequence the convergence holds for the whole sequence $\{\theta_k\}$, hence we obtain (2.3.8).

(c) follows directly from assumption (A2) and Lemma 2.3.1 (b). ∎

Lemma 2.3.3. *Suppose (A2),(A4) hold. Let $\hat{c} \in L^2(0,T;L^2(\Omega))$. Then there exists a unique $\theta(\hat{c}) \in W_2^{2,1}(Q_T)$ and a unique $z(\hat{c}) = (p(\hat{c}), m(\hat{c})) \in W^{1,\infty}(0,T;L^\infty(\Omega)) \times W^{1,\infty}(0,T;L^\infty(\Omega))$, satisfying*

$$\rho\alpha(\theta)\frac{\partial \theta}{\partial t} - k\Delta\theta = B(\theta,\hat{c}) \quad \text{in } Q_T \tag{2.3.10a}$$

$$-k\frac{\partial \theta}{\partial \nu} = h(\theta - \theta_\Gamma) \quad \text{on } \partial\Omega \times (0,T) \tag{2.3.10b}$$

$$\theta(x,0) = \theta_0 \quad \text{in } \Omega \tag{2.3.10c}$$

$$z_t = f(z,\theta,\hat{c}) \quad \text{in } Q_T \tag{2.3.10d}$$

$$z(0) = 0 \quad \text{in } \Omega. \tag{2.3.10e}$$

where f is defined as in (2.3.1a). Moreover, there exist $\lambda_1, \lambda_2 > 0$ such that

$$\|\theta_1 - \theta_2\|_{L^2(0,t;L^2(\Omega))}^2 \leq \lambda_1 \int_0^t \|\hat{c}_1 - \hat{c}_2\|_{L^2(0,s;L^2(\Omega))}^2 ds \tag{2.3.11}$$

and

$$\||z_1 - z_2|\|_{L^2(0,t;L^2(\Omega))}^2 \leq \lambda_2 \int_0^t \|\hat{c}_1 - \hat{c}_2\|_{L^2(0,s;L^2(\Omega))}^2 ds, \tag{2.3.12}$$

where (θ_i, z_i) is the solution corresponding to $\hat{c}_i$, $i = 1, 2$.

Proof of Lemma 2.3.3 Existence. We introduce the operator

$$P : L^2(Q_T) \to L^2(Q_T),$$

$$\theta = P\hat{\theta},$$

by demanding θ to be the solution of the linear parabolic problem

$$\rho\alpha(\hat{\theta})\frac{\partial \theta}{\partial t} - k\Delta\theta = B(\hat{\theta},\hat{c}) \quad \text{in } Q_T \tag{2.3.13a}$$

$$-k\frac{\partial \theta}{\partial \nu} = h(\theta - \theta_\Gamma) \quad \text{on } \partial\Omega \times (0,T) \tag{2.3.13b}$$

$$\theta(x,0) = \theta_0 \quad \text{in } \Omega. \tag{2.3.13c}$$

2.3. PROOF OF EXISTENCE

According to classical results about parabolic equations, problem (2.3.13) has a unique strong solution $\theta \in W_2^{1,1}(Q_T)$ (see [33][Chap. 3, Theorem 6.1- Remark 6.2]), therefore the operator P is well-defined. Further, applying classical regularity results for elliptic equations it yields $\theta \in W_2^{2,1}(Q_T)$. Moreover, thanks to Lemma 2.3.2 (a), there exists a constant $M > 0$, independent of $\hat{\theta}$, such that:

$$\|\theta\|_{W_2^{2,1}(Q_T)} \leq M. \tag{2.3.14}$$

We shall now show the continuity of the operator P.

Let $(\hat{\theta}_n) \subset L^2(Q_T)$ with $\hat{\theta}_n \to \hat{\theta}$ strongly in $L^2(Q_T)$. Defining $\theta_n = P\hat{\theta}_n$, in view of (2.3.14), $\|\theta_n\|_{W_2^{2,1}(Q_T)} \leq M$. Thus, we can find a sub-sequence $\hat{\theta}_{n'}$ such that

$$\theta_{n'} \to \theta \quad \text{weakly in} \quad W_2^{2,1}(Q_T), \quad \text{strongly in } L^2(Q_T), \tag{2.3.15a}$$

$$\theta_{n'} \to \theta \quad \text{a.e. in } Q_T. \tag{2.3.15b}$$

Testing equation (2.3.13a), written for the index n', by $\phi \in L^2(0,t;W_2^1(\Omega))$, we get

$$\int_0^t \int_\Omega \rho \alpha(\hat{\theta}_{n'})\frac{\partial \theta_{n'}}{\partial s}\phi\,dx\,ds + k\int_0^t \int_\Omega \nabla \theta_{n'} \nabla \phi\,dx\,ds$$
$$+ \int_0^t \int_{\partial\Omega} h(\sigma)(\theta_{n'} - \theta_\Gamma)\phi\,d\sigma\,ds - \int_0^t \int_\Omega B(\hat{\theta}_{n'},\hat{c})\phi\,dx\,ds = 0. \tag{2.3.16}$$

By means of (2.3.15a), (2.3.15b) we can pass to the limit in the last three terms of (2.3.16). We can break the first term in two terms

$$\rho\int_0^t\int_\Omega \alpha(\hat{\theta}_{n'})\frac{\partial \theta_{n'}}{\partial s}\phi\,dx\,ds = \rho\int_0^t\int_\Omega \alpha(\hat{\theta}_{n'})(\frac{\partial \theta_{n'}}{\partial s} - \frac{\partial \theta}{\partial s})\phi\,dx\,ds + \rho\int_0^t\int_\Omega \alpha(\hat{\theta}_{n'})\frac{\partial \theta}{\partial s}\phi\,dx\,ds.$$

Thanks to the continuity of α, we have that

$$\alpha(\hat{\theta}_{n'})\phi \to \alpha(\hat{\theta})\phi \quad \text{a.e. in } Q_T$$

thus, using Lebesgue theorem, $\rho\alpha(\hat{\theta}_{n'})\phi \to \rho\alpha(\hat{\theta})\phi$ strongly in $L^2(Q_T)$ while $\frac{\partial \theta_{n'}}{\partial s} \to \frac{\partial \theta}{\partial s}$ weakly in $L^2(Q_T)$. Thus, $\int_0^t\int_\Omega \alpha(\hat{\theta}_{n'})(\frac{\partial \theta_{n'}}{\partial s} - \frac{\partial \theta}{\partial s})\phi\,dx\,ds \to 0$ and

$$\rho\int_0^t\int_\Omega \alpha(\hat{\theta}_{n'})\frac{\partial \theta_{n'}}{\partial s}\phi\,dx\,ds \to \rho\int_0^t\int_\Omega \alpha(\hat{\theta})\frac{\partial \theta}{\partial s}\phi\,dx\,ds.$$

Hence we have obtained

$$\rho \int_0^t \int_\Omega \alpha(\hat{\theta})\theta_s \phi \, dx \, ds + k \int_0^t \int_\Omega \nabla\theta \nabla\phi \, dx \, ds$$
$$+ \int_0^t \int_{\partial\Omega} h(\sigma)(\theta - \theta_\Gamma)\phi \, d\sigma \, ds - \int_0^t \int_\Omega B(\hat{\theta}, \hat{c})\phi \, dx \, ds = 0.$$

As the solution to the parabolic system (2.3.13) is unique, we have

$$\theta = P\hat{\theta} \quad \text{a.e. in } Q_T$$

and, since the limit does not depend on the extracted sub-sequence, it follows that

$$P\hat{\theta}_n \to P\hat{\theta}$$

weakly in $W_2^{2,1}(Q_T)$ and strongly in $L^2(Q_T)$.

Now, let

$$K := \{u \in L^2(Q_T) : \|u\|_{W_2^{2,1}(Q_T)} \leq M\}.$$

K is non-empty, convex, closed and relatively compact subset of $L^2(Q_T)$ and $P : K \subset L^2(Q_T) \to K$ is a continuous mapping. By the Schauder fixed point theorem, there exists a fixed point of the mapping P, i.e. there exists a weak solution $\theta \in W_2^{2,1}(Q_T)$ to (2.3.10a),(2.3.10b),(2.3.10c). Since the equation $z_t = f(z, \theta, \hat{c})$ with initial condition has already been solved and incorporated in the right-hand side $B(\theta, \hat{c})$, the proof provides actually the existence of a solution for the whole system (2.3.10a)-(2.3.10e). Now we must prove (2.3.11), which can be seen as a kind of stability of the solution θ with respect to given θ and c.

Uniqueness and stability. Let

$$J(\theta) := \int_0^\theta \rho\alpha(\xi) d\xi. \tag{2.3.17}$$

Integration of (1.4.6) with respect to time leads to

$$\int_0^t B(\theta, c)(x, s) ds = J(\theta(x,t)) - J(\theta_0(x)) - k\Delta \int_0^t \theta(x, s) ds. \tag{2.3.18}$$

Now, let $\theta_1, \theta_2 \in W_2^{2,1}(Q_T)$ be solutions to the system (2.3.13) corresponding to $\hat{c}_1, \hat{c}_2$

2.3. PROOF OF EXISTENCE

respectively. Inserting these solutions into (2.3.18), subtracting both equations, and testing by $\theta := \theta_1 - \theta_2$, we find

$$\int_0^t \int_\Omega \Big(\int_0^s B(\theta_1(x,\xi), \hat{c}_1(x,\xi)) - B(\theta_2(x,\xi), \hat{c}_2(x,\xi)) d\xi \Big) \theta(x,s) dx\, ds$$

$$= \int_0^t \int_\Omega [J(\theta_1(x,s)) - J(\theta_2(x,s))] \theta(x,s) dx\, ds + k \int_0^t \int_\Omega \nabla \Big(\int_0^s \theta(x,\xi) d\xi \Big) \nabla \theta(x,s) dx\, ds$$

$$+ \int_0^t \int_{\partial\Omega} \Big(\int_0^s h(\sigma) \theta(\sigma,\xi) d\xi \Big) \theta(\sigma,s) d\sigma\, ds. \qquad (2.3.19)$$

Concerning the last term we can see that

$$\int_0^t \int_{\partial\Omega} \Big(\int_0^s h(\sigma) \theta(\sigma,\xi) d\xi \Big) \theta(\sigma,s) d\sigma\, ds = \int_0^t \int_{\partial\Omega} h(\sigma) \Big(\int_0^s \theta(\sigma,\xi) d\xi \Big) \theta(\sigma,s) d\sigma ds$$

$$= \frac{1}{2} \int_0^t \int_{\partial\Omega} h(\sigma) \frac{d}{ds} \Big(\int_0^s \theta(\sigma,\xi) d\xi \Big)^2 d\sigma ds = \frac{1}{2} \int_{\partial\Omega} h(\sigma) \Big(\int_0^t \theta(\sigma,s) ds \Big)^2 d\sigma.$$

Thus, from (2.3.19) we obtain

$$\int_0^t \int_\Omega \Big(\int_0^s B(\theta_1(x,\xi), \hat{c}_1(x,\xi)) - B(\theta_2(x,\xi), \hat{c}_2(x,\xi))(x,\xi)) d\xi \Big) \theta(x,s) dx\, ds$$

$$\geq \alpha_0 \rho \int_0^t \int_\Omega \theta^2(x,s) dx\, ds + \frac{k}{2} \int_\Omega \Big| \nabla \int_0^t \theta(x,s) ds \Big|^2 dx$$

$$+ \frac{1}{2} \int_{\partial\Omega} h(\sigma) \Big(\int_0^t \theta(\sigma,s) ds \Big)^2 d\sigma \geq \alpha_0 \rho \int_0^t \int_\Omega \theta^2(x,s) dx\, ds.$$

Using Holder's and Young's inequalities and Lemma 2.3.1 it follows that

$$\Big| \int_0^t \int_\Omega \Big(\int_0^s B(\theta_1(x,\xi), \hat{c}_1(x,\xi)) - B(\theta_2(x,\xi), \hat{c}_2(x,\xi)) d\xi \Big) \theta(x,s) dx\, ds \Big|$$

$$\leq \frac{1}{4\delta} \int_0^t \int_\Omega \Big(\int_0^s B(\theta_1(x,\xi), \hat{c}_1(x,\xi)) - B(\theta_2(x,\xi), \hat{c}_2(x,\xi)) d\xi \Big)^2 dx\, ds$$

$$+ \delta \int_0^t \|\theta(\cdot,s)\|_{L^2(\Omega)}^2 ds$$

$$\leq \frac{T}{4\delta} \int_0^t \int_\Omega \int_0^s \left(K_1 |\theta_1(x,\xi) - \theta_2(x,\xi)|^2 + K_2 |\hat{c}_1(x,\xi) - \hat{c}_2(x,\xi)|^2 \right) d\xi\, dx\, ds$$

$$+ \delta \int_0^t \|\theta(\cdot, s)\|_{L^2(\Omega)}^2 \, ds$$

$$\leq \frac{CT}{4\delta} \int_0^t \|\theta\|_{L^2(0,s;L^2(\Omega))}^2 \, ds + \frac{CT}{4\delta} \int_0^t \|\hat{c}\|_{L^2(0,s;L^2(\Omega))}^2 \, ds + \delta \int_0^t \|\theta(\cdot, s)\|_{L^2(\Omega)}^2 \, ds,$$

where $C = \min\{K_1, K_2\}$. Thus, we have

$$\alpha_0 \rho \int_0^t \int_\Omega \theta^2(x,s) dx\, ds \leq \frac{CT}{4\delta} \int_0^t \|\theta\|_{L^2(0,s;L^2(\Omega))}^2 ds$$

$$+ \frac{CT}{4\delta} \int_0^t \|\hat{c}\|_{L^2(0,s;L^2(\Omega))}^2 \, ds + \delta \int_0^t \|\theta(\cdot, s)\|_{L^2(\Omega)}^2 \, ds.$$

Choosing $\delta = \frac{\alpha_0 \rho}{2}$ we have:

$$\|\theta\|_{L^2(0,t;L^2(\Omega))}^2 \leq M_1 \int_0^t \|\theta\|_{L^2(0,s;L^2(\Omega))}^2 ds + M_2 \int_0^t \|\hat{c}\|_{L^2(0,s;L^2(\Omega))}^2 ds$$

with constants $M_1, M_2 > 0$.

Hence, applying Gronwall lemma, we find a constant C_1 such that

$$\|\theta_1 - \theta_2\|_{L^2(0,t;L^2(\Omega))}^2 \leq C_1 \int_0^t \|\hat{c}_1(s) - \hat{c}_2(s)\|_{L^2(0,s;L^2(\Omega))}^2 ds. \tag{2.3.20}$$

Therefore (2.3.11) is proved. Inequality (2.3.12) follows immediately from Lemma 2.3.1 (b) and estimate (2.3.11). The proof of Lemma 2.3.3 is thus completed.

∎

Now, we are in a position to prove Theorem 2.2.1.

Proof of Theorem 2.2.1
Let us denote
$$\mu(\theta, c) := (1 - p - m) D(\theta, c).$$

We note that, in view of (A6) and Lemma 2.3.1 (b), μ is Lipschitz-continuous with respect to θ and c.

2.3. PROOF OF EXISTENCE

We define an operator
$$\mathcal{T} : L^2(Q_T) \longrightarrow L^2(Q_T),$$

$$\mathcal{T}\hat{c} = c, \tag{2.3.21}$$

by demanding c to be the solution of the parabolic problem

$$\frac{\partial c}{\partial t} - \operatorname{div}(\mu_{\hat{c}} \nabla c) = 0 \quad \text{in } Q_T \tag{2.3.22a}$$

$$-\mu_{\hat{c}} \frac{\partial c}{\partial \nu} = \beta(c - c_p) \quad \text{on } \partial\Omega \times (0, T) \tag{2.3.22b}$$

$$c(x, 0) = c_0 \quad \text{in } \Omega. \tag{2.3.22c}$$

where $\mu_{\hat{c}} = (1 - p_{\hat{c}} - m_{\hat{c}})D(\theta_{\hat{c}}, \hat{c})$, $(\theta_{\hat{c}}, p_{\hat{c}}, m_{\hat{c}})$ being the solution to problem (2.3.10), with respect to given $\hat{c}$. Denoting

$$a(c, \phi; t) := \int_\Omega \mu_{\hat{c}} \nabla c \, \nabla \phi \, dx + \int_{\partial\Omega} \beta \, c \, \phi \, d\sigma,$$

$$\langle f(t), \phi \rangle := \int_{\partial\Omega} \beta c_p \phi \, d\sigma, \qquad \phi \in W_2^1(\Omega)$$

we have that problem (2.3.22) is equivalent to the following one. We seek a function c such that, for all $\phi \in W_2^1(\Omega)$ and a.e in $t \in (0, T)$

$$\left\langle \frac{d}{dt} c(t), \phi \right\rangle + a(c(t), \phi; t) = \langle f(t), \phi \rangle, \tag{2.3.23a}$$

$$c(0) = c_0, \tag{2.3.23b}$$

$$c \in W(0, T), \tag{2.3.23c}$$

where $\langle \, , \, \rangle$ denotes the duality between $W_2^1(\Omega)$ and $(W_2^1(\Omega))^*$.

In view of (A3),(A6),(A7), problem (2.3.23) admits a unique solution $c \in W(0, T)$ (cf. [36, Theorem 1.2, Chap. 2]). Moreover, there exists a constant M independent of $\hat{c}$, such that:

$$\|c\|_{W(0,T)} \leq M. \tag{2.3.24}$$

To derive the continuity of the operator $\mathcal{T}$, let $\{\hat{c}_n\} \subset L^2(0, T; L^2(\Omega))$, with $\hat{c}_n \to \hat{c}$ strongly in $L^2(Q_T)$. Defining $c_n = \mathcal{T}\hat{c}_n$, thanks to (2.3.24), we have $\|c_n\|_{W(0,T)} \leq M$.

Thus, there exists a sub-sequence $\{\hat{c}_{n'}\}$ such that

$$c_{n'} \longrightarrow c \quad \text{weakly in } W(0,T). \tag{2.3.25}$$

We test (2.3.22a) by

$$\Phi(x,t) = \psi(t)\phi(x) \quad \text{with} \quad \psi \in C^1[0,T], \ \psi(T) = 0, \ \phi \in W_2^1(\Omega). \tag{2.3.26}$$

Denoting $T\hat{c}_{n'} := c_{n'}$, we have

$$\int_0^T \int_\Omega \frac{\partial c_{n'}}{\partial s} \Phi \, dx \, ds + \int_0^T \int_\Omega \mu_{\hat{c}_{n'}} \nabla c_{n'} \nabla \Phi \, dx \, ds + \int_0^T \int_{\partial\Omega} \beta(c_{n'} - c_p) \Phi \, d\sigma ds = 0. \tag{2.3.27}$$

Concerning the first term in (2.3.27) we have

$$\int_0^T \int_\Omega \frac{\partial c_{n'}}{\partial s} \Phi \, dx \, ds = -\int_\Omega c_{n'}(x,0) \, \Phi(x,0) dx \; - \int_0^T \int_\Omega c_{n'} \frac{\partial \Phi}{\partial s} \, dx \, ds.$$

Now,

$$\int_\Omega c_{n'}(x,0) \, \Phi(x,0) dx = \int_\Omega c_0 \, \Phi(x,0) dx,$$

and, by virtue of (2.3.25),

$$\int_0^T \int_\Omega c_{n'} \frac{\partial \Phi}{\partial s} \, dx \, ds \to \int_0^T \int_\Omega c \frac{\partial \Phi}{\partial s} \, dx \, ds \;.$$

The second term can be rearranged as

$$\int_0^T \int_\Omega \mu_{\hat{c}_{n'}} \nabla c_{n'} \nabla \Phi \, dx \, ds = \int_0^T \int_\Omega \mu_{\hat{c}_{n'}} (\nabla c_{n'} - \nabla c) \nabla \Phi \, dx \, ds + \int_0^T \int_\Omega \mu_{\hat{c}_{n'}} \nabla c \nabla \Phi \, dx \, ds.$$

Since μ is continuous and bounded as a function of c, possibly extracting a subsequence, we obtain: $\mu_{\hat{c}_{n'}}(x,t) \to \mu_{\hat{c}}(x,t)$ a.e in Q_T, therefore $\mu_{\hat{c}_{n'}}(x,t)\nabla\Phi \to \mu_{\hat{c}}(x,t)\nabla\Phi$ pointwise, moreover $\mu_{\hat{c}_{n'}}(x,t)\nabla\Phi$ is bounded in L^2 thus, using Lebesgue theorem, we have

$$\mu_{\hat{c}_{n'}} \nabla\Phi \to \mu_{\hat{c}} \nabla\Phi \quad \text{strongly} \quad \text{in} \quad L^2(Q_T).$$

2.3. PROOF OF EXISTENCE

Moreover, $(\nabla c_{n'} - \nabla c) \to 0$ weakly in $L^2(Q_T)$ because of (2.3.25), thus we obtain

$$\int_0^T \int_\Omega \mu_{\hat{c}_{n'}} \nabla c_{n'} \nabla \Phi \, dx \, ds \to \int_0^T \int_\Omega \mu_{\hat{c}} \nabla c \nabla \Phi \, dx \, ds.$$

Applying the trace theorem, the last term in (2.3.27) converges too. Thus, we can pass to the limit in (2.3.27) obtaining

$$-\psi(0) \int_\Omega c_0 \, \phi(x) \, dx - \int_0^T \int_\Omega c \psi_s \phi \, dx \, ds$$
$$+ \int_0^T \psi \int_\Omega \mu_{\hat{c}} \nabla c \nabla \phi \, dx \, ds + \int_0^T \psi \int_{\partial\Omega} \beta(c - c_p) \phi \, d\sigma \, ds = 0. \quad (2.3.28)$$

Consequently

$$\int_0^T \psi \Big(\int_\Omega c_s \phi \, dx + \int_\Omega \mu_{\hat{c}} \nabla c \nabla \phi \, dx + \int_{\partial\Omega} \beta(c - c_p) \phi \, d\sigma \Big) ds = 0.$$

The above is true for ϕ, ψ satisfying (2.3.28). Therefore (2.3.28) gives, a.e in $t \in (0, T)$

$$\Big\langle \frac{d}{dt} c(t), \phi \Big\rangle + a(t; c(t), \phi) = \langle F(t), \phi \rangle \quad \forall \phi \in W_2^1(\Omega).$$

Since the solution of (2.3.22) is unique, we can conclude

$$\mathcal{T}\hat{c} = c,$$

and, since the limit does not depend on the extracted sub-sequence, it follows that

$$\mathcal{T}\hat{c}_n \to \mathcal{T}c \quad (2.3.29)$$

weakly in $W(0, T)$ and strongly in $L^2(Q_T)$.

Now, let

$$K := \{ v \in L^2(Q_T) : \|v\|_{W(0,T)} \leq M \}.$$

K is convex and compact in $L^2(Q_T)$ and $\mathcal{T} : K \subset L^2(Q_T) \to K$ is a continuous mapping. By the Schauder fixed point theorem the proof of the existence of solutions is concluded. ■

CHAPTER 2. ANALYSIS OF THE COMPLETE MODEL

Proof of Proposition 2.2.2

We prove now that $\theta \in L^\infty(Q_T)$.

To this end, we write the weak form of equation (2.1.1):

$$\int_0^t \left\langle \rho\alpha \frac{\partial \theta}{\partial t}, \phi \right\rangle ds + \int_0^t \int_\Omega k \nabla\theta \nabla\phi \, dx \, ds + \int_0^t \int_{\partial\Omega} h(\theta - \theta_\Gamma)\phi \, d\sigma \, ds \qquad (2.3.30)$$
$$= \int_0^t \langle B(\theta,c)\phi \rangle \, ds.$$

Testing with the function $\phi = (\theta - \theta_\Gamma)^- = -\min\{\theta - \theta_\Gamma, 0\}$, through integration by parts, we obtain from the first term:

$$\int_0^t \rho \left\langle \alpha \frac{\partial \theta}{\partial t}, \phi \right\rangle ds \geq \rho\alpha_1 \int_0^t \left\langle \frac{\partial \theta}{\partial t}, \phi \right\rangle ds$$
$$= \rho\alpha_1 \int_0^t \left\langle \frac{\partial (\theta - \theta_\Gamma)}{\partial t}, \phi \right\rangle ds = \rho\alpha_1 \frac{1}{2} \int_\Omega |\phi(t)|^2 dx, \quad (2.3.31)$$

from the second term

$$k \int_0^t \int_\Omega \nabla\theta \nabla\phi \, dx \, ds = \int_0^t \int_{\{x \in \Omega: \theta - \theta_\Gamma \leq 0\}} \nabla\phi \nabla\phi \, dx \, ds$$
$$= \int_0^t \int_\Omega |\nabla\phi|^2 \, dx \, ds \geq 0, \quad (2.3.32)$$

from the third term

$$\int_0^t \int_{\partial\Omega} h(\theta - \theta_\Gamma)\phi \, d\sigma \, ds = \int_0^t \int_{\{\sigma \in \partial\Omega: \theta - \theta_\Gamma \leq 0\}} h(\theta - \theta_\Gamma)^2 \, d\sigma \, ds \geq 0. \quad (2.3.33)$$

Regarding the right-hand side of (2.3.30), we notice that if $(\theta - \theta_\Gamma)^- \neq 0$, then $\theta \leq \theta_\Gamma$, but in this range of temperature we know that $B(\theta,c) = 0$, since $g_i = 0$ and vice-versa: if $B(\theta,c) \neq 0$, then $(\theta - \theta_\Gamma)^- = 0$. Therefore the right-hand side is always zero and we can conclude, from (2.3.30)-(2.3.33), that:

$$\frac{1}{2}\int_\Omega |(\theta - \theta_\Gamma)^-(t)|^2 dx \leq 0 \qquad \forall t \in (0,T),$$

2.3. PROOF OF EXISTENCE

thus

$$(\theta - \theta_\Gamma)^-(t) = 0 \quad \forall t \in (0,T) \quad \text{a.e. in } \Omega$$

and $\theta(x,t) \geq \theta_\Gamma$, $\forall t \in (0,T)$.
A similar argument, using the test function $\phi = (\theta - \theta_0)^+$, leads to $\theta(x,t) \leq \theta_0(x)$ a.e. in Q_T.

Now we pass to consider the weak form of equation (2.1.2):

$$\int_0^t \left\langle \frac{\partial c}{\partial t}, \phi \right\rangle ds + \int_0^t \int_\Omega \mu_{\theta,c} \nabla c \nabla \phi \, dx \, ds + \int_0^t \int_{\partial \Omega} \beta(c - c_p)\phi \, d\sigma ds = 0, \quad (2.3.34)$$

$\forall \phi \in L^2(0,t;W_2^1(\Omega))$ and for every $t \in (0,T)$, where $\langle , \rangle$ denotes the duality pairing between $W_2^1(\Omega)$ and $(W_2^1(\Omega))^*$.
Testing with the function $\phi = (c - C_1)^- = -\min\{c - C_1, 0\}$, through integration by parts, we get from the first term:

$$\int_0^t \left\langle \frac{\partial c}{\partial t}, \phi \right\rangle ds = \int_0^t \left\langle \frac{\partial (c - C_1)}{\partial t}, \phi \right\rangle ds = \frac{1}{2} \int_\Omega |\phi(t)|^2 dx, \quad (2.3.35)$$

from the second term

$$\int_0^t \int_\Omega \mu_{\theta,c} \nabla c \nabla \phi \, dx \, ds = \int_0^t \int_{\{x \in \Omega: c - C_1 \leq 0\}} \mu_{\theta,c} \nabla \phi \nabla \phi \, dx ds$$

$$= \int_0^t \int_\Omega \mu_{\theta,c} |\nabla \phi|^2 \, dx ds > 0 \quad (2.3.36)$$

and from the third term

$$\int_0^t \int_{\partial \Omega} \beta(c - c_p)\phi \, d\sigma ds = \int_0^t \int_{\{\sigma \in \partial\Omega: c - C_1 \leq 0\}} \beta(c - c_p)(c - C_1) \, d\sigma ds$$

$$\geq \int_0^t \int_{\{\sigma \in \partial\Omega: c - C_1 \leq 0\}} \beta(c - C_1)^2 \, d\sigma ds \geq 0 \quad (2.3.37)$$

Thus, from (2.3.34)-(2.3.37), we obtain

$$\frac{1}{2} \int_\Omega |(c - C_1)^-(t)|^2 dx \leq 0 \quad \forall t \in (0,T),$$

thus
$$(c - C_1)^-(t) = 0 \quad \forall t \in (0,T) \quad \text{a.e. in } \Omega$$

and $c(x,t) \geq C_1$, $\forall t \in (0,T)$. A similar argument, using the test function $\phi = (c-C_2)^+$, leads to $c(x,t) \leq C_2$ a.e in Q_T. ∎

2.4 Proof of uniqueness

We commence with the following regularity result:

Lemma 2.4.1. *Under the assumptions of Theorem 2.2.3, the solutions θ, c to the initial-boundary values problems related to equations (2.1.1)–(2.1.2) are in $W_5^{2,1}(Q_T)$.*

Proof. Until now we have proved the existence of at least one solution for the initial-boundary value problems related to equations (2.1.1), (2.1.2) and we proved that the solutions are bounded in $L^\infty(Q_T)$ (Prop. 2.2.2). Then we can directly apply a classical result from Ladyženskaja ([33, Theorem 10.1, Chap. 3]), which gives us that θ and c are Hölder continuous functions, with Hölder exponent depending, among the others, also on the maximum of the functions over Q_T. It follows that the right-hand sides of the ODEs (2.1.3)–(2.1.4) are continuous functions, therefore the corresponding solutions, p and m, are continuously differentiable.

Thus, the PDEs (2.1.1)–(2.1.2) have continuous coefficients and, since $\theta_0, c_0 \in W_5^2(\Omega)$ and $h, \beta \in W_5^1(\partial\Omega)$, we can apply another classical result from Ladyženskaja ([33, Theorem 9.1, Chap. 4]) which yields: $\theta, c \in W_5^{2,1}(Q_T)$. ∎

Lemma 2.4.2. *Assuming that α is constant, we have that, for every $c_1, c_2 \in L^2(Q_T)$, there exists a constant $M > 0$ such that, for the corresponding θ_1, θ_2, it holds:*

$$\|\theta_1 - \theta_2\|^2_{W_2^{2,1}(Q_T)} \leq M \|c_1 - c_2\|^2_{L^2(Q_T)}. \tag{2.4.1}$$

Proof.

We consider the heat equation of our system:

$$\rho\alpha \frac{\partial \theta}{\partial t} = k\Delta\theta + \rho L_p p_t + \rho L_m m_t. \tag{2.4.2}$$

We write (2.4.2) for θ_1, c_1, p_1, m_1 and θ_2, c_2, p_2, m_2. Subtracting and denoting as usual

2.4. PROOF OF UNIQUENESS

$\theta = \theta_1 - \theta_2$, we see that the difference satisfies the following system:

$$\rho\alpha\frac{\partial\theta}{\partial t} - k\Delta\theta = \rho(L_p(\theta_1)p_{1,t} - L_p(\theta_2)p_{2,t}) + \rho(L_m(\theta_1)m_{1,t} - L_m(\theta_2)m_{2,t}) =: f$$
$$-k\frac{\partial\theta}{\partial \nu} = h\theta$$
$$\theta(x,0) = 0.$$

Applying again standard parabolic theory (cf.[37, Theorem 6.2, Chap. 4]), we know that there exists a positive constant K such that we can estimate the norm of the solution as follows:

$$\|\theta\|_{W_2^{2,1}(Q_T)} \leq K\|f\|_{L^2(Q_T)}.$$

Now we can estimate the term $\|f\|_{L^2(Q_T)}$, by means of Lemma 2.3.1 (b) and assumptions (A2), as:

$$\|\rho(L_p(\theta_1)p_{1,t} - L_p(\theta_2)p_{2,t}) + \rho(L_m(\theta_1)m_{1,t} - L_m(\theta_2)m_{2,t})\|_{L^2(Q_T)}$$
$$\leq K_1\|\theta_1 - \theta_2\|_{L^2(Q_T)} + K_2\|c_1 - c_2\|_{L^2(Q_T)},$$

with K_1, K_2 positive constants. Now, using Lemma 2.3.3, we can estimate the term $\|\theta_1 - \theta_2\|_{L^2(Q_T)}$ and therewith finish the proof. ∎

Lemma 2.4.3. Let $u \in L^\infty(0,T;L^2(\Omega)) \cap L^2(0,T;W_2^1(\Omega))$, then there holds

$$\int_0^T \|u(t)\|_{L^{10/3}(\Omega)}^{10/3} dt \leq \left(\int_0^T \|u(t)\|_{L^6(\Omega)}^2 dt\right) \|u\|_{L^\infty(0,T;L^2(\Omega))}^{4/3}.$$

Proof. Owing to Riesz' convexity theorem (cf. [8, Oss. 2, Chap. 4]), we have

$$\|u\|_{L^r(\Omega)} \leq \|u\|_{L^{q_1}(\Omega)}^{1-\Theta} \|u\|_{L^{q_2}(\Omega)}^{\Theta},$$

for all $u \in L^{q_1}(\Omega) \cap L^{q_2}(\Omega)$ with $1 \leq q_1, q_2 < \infty$, $0 < \Theta < 1$, and $\frac{1}{r} = \frac{1-\Theta}{q_1} + \frac{\Theta}{q_2}$. Invoking the continuous embedding $W_2^1(\Omega) \subset L^6(\Omega)$, the assertion follows by defining $q_1 = 6$, $q_2 = 2$, $\Theta = \frac{2}{5}$, and $r = \frac{10}{3}$. ∎

We are now in a position to prove Theorem 2.2.3.

Proof of Theorem 2.2.3

We write equation (2.1.2) for c_1 and c_2, subtract, integrate over Q_T and test by $c_1 - c_2$. In the sequel we will use the following notations: $c = c_1 - c_2$, $\theta = \theta_1 - \theta_2$, $p =$

$p_1 - p_2$, $m = m_1 - m_2$.
We have:

$$\frac{1}{2}\int_\Omega c^2(t)dx + \int_0^t\int_\Omega \Big((1-p_1-m_1)D(\theta_1)\nabla c_1 - (1-p_2-m_2)D(\theta_2)\nabla c_2\Big)\nabla c\,dxds$$
$$+ \int_0^t\int_{\partial\Omega} \beta c^2 d\sigma ds = 0.$$

Now,

$$\int_0^t\int_\Omega \Big((1-p_1-m_1)D(\theta_1)\nabla c_1 - (1-p_2-m_2)D(\theta_2)\nabla c_2\Big)\nabla c\,dxds$$
$$= \int_0^t\int_\Omega (1-p_1-m_1)D(\theta_1)|\nabla c|^2 dxds - \int_0^t\int_\Omega (p+m)D(\theta_1)\nabla c_2\nabla c\,dxds$$
$$+ \int_0^t\int_\Omega (1-p_2-m_2)(D(\theta_1)-D(\theta_2))\nabla c_2\nabla c\,dxds. \qquad (2.4.3)$$

Denoting, in the sequel, by K_i generic positive constants independent of θ and c, we obtain:

$$\frac{1}{2}\int_\Omega c^2(t)dx + K_1\int_0^t \|\nabla c\|^2_{L^2(\Omega)} dxds$$
$$\leq \int_0^t\int_\Omega |p+m||D(\theta_1)||\nabla c_2||\nabla c|\,dxds$$
$$+ \int_0^t\int_\Omega |1-p_2-m_2||D(\theta_1)-D(\theta_2)||\nabla c_2||\nabla c|\,dxds. \qquad (2.4.4)$$

By means of Lemma 2.4.1, we know that $c_2 \in W_5^{2,1}(Q_T)$. According to Amann (cf [1, Theorem 1.1], Theorem 1.1), we have the embedding $W_5^{2,1}(Q_T) \hookrightarrow C([0,T]; W_5^1(\Omega))$.

2.4. PROOF OF UNIQUENESS

Thus, we can estimate the first term in the right hand-side of (2.4.4) as:

$$\int_0^t \int_\Omega |p+m| |D(\theta_1)| |\nabla c_2| |\nabla c| \, dx ds$$
$$\leq \int_0^t \|p+m\|_{L^{10/3}(\Omega)} \|\nabla c_2\|_{L^5(\Omega)} \|D(\theta_1)\|_{L^\infty(\Omega)} \|\nabla c\|_{L^2(\Omega)} ds \quad (2.4.5)$$
$$\leq \frac{K_1}{4} \int_0^t \|\nabla c\|_{L^2(\Omega)}^2 ds + \frac{4K_2}{K_1} \int_0^t \|p+m\|_{L^{10/3}(\Omega)}^2 ds.$$

Moreover, thanks to Lemma 2.3.1 (b), we get:

$$\int_0^t \|p+m\|_{L^{10/3}(\Omega)}^2 ds = \int_0^t \left[\int_\Omega |p+m|^{10/3} dx\right]^{3/5} ds$$
$$\leq K_3 \int_0^t \left[\int_0^s \int_\Omega \theta^{10/3} dx d\tau\right]^{3/5} ds + K_4 \int_0^t \left[\int_0^s \int_\Omega c^{10/3} dx d\tau\right]^{3/5} ds. \quad (2.4.6)$$

Now, we apply Lemma 2.4.3, Young's inequality and the embedding $W_2^1(\Omega) \hookrightarrow L^6(\Omega)$ to the right-hand side of (2.4.6), obtaining:

$$\int_0^t \left[\int_0^s \int_\Omega \theta^{10/3} dx d\tau\right]^{3/5} ds \leq \int_0^t \left[\int_0^s K_5 \|\theta\|_{W_2^1(\Omega)}^2 d\tau\right]^{3/5} \|\theta\|_{L^\infty(0,s;L^2(\Omega))}^{4/5} ds$$
$$\leq \frac{3K_5}{5} \int_0^t \int_0^s \|\theta\|_{W_2^1(\Omega)}^2 d\tau ds + \frac{2}{5} \int_0^t \|\theta\|_{L^\infty(0,s;L^2(\Omega))}^2 ds. \quad (2.4.7)$$

Analogously, it holds:

$$\int_0^t \left[\int_0^s \int_\Omega c^{10/3} dx d\tau\right]^{3/5} ds \leq K_5 \frac{3}{5} \int_0^t \int_0^s \|c\|_{W_2^1(\Omega)}^2 d\tau ds + \frac{2}{5} \int_0^t \|c\|_{L^\infty(0,s;L^2(\Omega))}^2 ds. \quad (2.4.8)$$

Regarding the second term in the right-hand side of (2.4.4), using Lemma 4.1, assumption (A6) and Young's inequality again, we have:

$$\int_0^t \int_\Omega |1-p_2-m_2| |D(\theta_1) - D(\theta_2)| |\nabla c_2| |\nabla c| \, dx ds \leq \|\nabla c_2\|_{L^\infty(0,t;L^3(\Omega))} \int_0^t \|\theta\|_{L^6(\Omega)} \|\nabla c\|_{L^2(\Omega)} ds$$
$$\leq K_6 \int_0^t \|\theta\|_{W_2^1(\Omega)}^2 ds + \frac{K_1}{4} \int_0^t \|\nabla c\|_{L^2(\Omega)}^2 ds. \quad (2.4.9)$$

Using (5.4) to (2.4.9), we find that

$$\min\left\{\frac{1}{2}, \frac{K_1}{2}\right\} \int_\Omega c^2(t)dx + \int_0^t \|\nabla c\|^2_{L^2(\Omega)}ds$$
$$\leq K_7 \left(\int_0^t \|\theta\|^2_{W_2^1(\Omega)}ds + \int_0^t \|\theta\|^2_{L^\infty(0,s;L^2(\Omega))}ds \right) + \frac{3K_5}{5} \int_0^t \int_0^s \|\theta\|^2_{W_2^1(\Omega)}d\tau ds \quad (2.4.10)$$
$$+ K_8 \left(\int_0^t \int_0^s \|c\|^2_{W_2^1(\Omega)}d\tau ds + \int_0^t \|c\|^2_{L^\infty(0,s;L^2(\Omega))}ds \right).$$

Thanks to the embedding $W_2^{2,1}(Q_T) \hookrightarrow C([0,T]; W_2^1(\Omega))$ (see [2], Theorem 1.1) and Lemma 5.2, we obtain for the first term in the right-hand side of (2.4.10):

$$\int_0^t \|\theta(s)\|^2_{W_2^1(\Omega)}ds \leq \int_0^t \|\theta\|^2_{L^\infty(0,s;W_2^1(\Omega))}ds \leq K_9 \int_0^t \|\theta\|_{W_2^{2,1}(Q_s)}ds \leq \int_0^t K_{10} \int_0^s \|c\|^2_{L^2(\Omega)}d\tau ds$$

Analogous estimates hold for the other terms involving θ. Thus we end up with:

$$\|c(t)\|^2_{L^2(\Omega)} + \int_0^t \|\nabla c\|^2_{L^2(\Omega)}ds \leq K_{11} \int_0^t \|c\|^2_{L^\infty(0,s;L^2(\Omega))}ds$$
$$+ K_{12} \int_0^t \int_0^s \|\nabla c\|^2_{L^2(\Omega)}d\tau ds + K_{13} \int_0^t \int_0^s \|c\|^2_{L^2(\Omega)}d\tau ds \quad \forall t \in [0,T].$$

Taking the essential supremum over $[0, \hat{t}]$ for some $\hat{t} \in [0,T]$ in the previous inequality, we obtain:

$$\|c(t)\|^2_{L^\infty(0,\hat{t}L^2(\Omega))} + \int_0^{\hat{t}} \|\nabla c\|^2_{L^2(\Omega)} \leq K_{11} \int_0^{\hat{t}} \|c\|^2_{L^\infty(0,s;L^2(\Omega))}ds$$
$$+ K_{12} \int_0^{\hat{t}} \int_0^s \|\nabla c\|^2_{L^2(\Omega)}d\tau ds + K_{13} \int_0^{\hat{t}} \int_0^s \|c\|^2_{L^2(\Omega)}d\tau ds.$$

Since $\int_0^{\hat{t}} \int_0^s \|c\|^2_{L^2(\Omega)}d\tau ds \leq T \int_0^{\hat{t}} \|c\|^2_{L^\infty(0,s;L^2(\Omega))}ds$, we can apply the Gronwall lemma and conclude the proof of Theorem 3.2. ∎

Chapter 3

Analysis of a related quasilinear parabolic system

3.1 Problem statement and strategy

The uniqueness result presented in the previous chapter has been obtained under quite strong assumptions. This leads us to consider a related system constituted of the two quasilinear parabolic equations, with the same structure of those present in the complete model, under weaker assumptions, not coupled though with the two additional ODEs for the phase transitions. It seems, in fact, that the analysis performed in this chapter can't be carried over to the whole system considered previously. Nevertheless we believe in the intrinsic mathematical interest of this chapter, dealing with a quite general and relevant class of parabolic systems, often encountered in problems arising from physics. Very similar systems of quasilinear parabolic equations are studied also for other applications (see for instance, [42] and [43]).

Here we examine, in the domain $Q_T = \Omega \times (0, T)$, $\Omega \subset \mathbb{R}^N$, the following system of equations:

$$\theta_t - \Delta \theta = r(\theta, c) \tag{3.1.1}$$

$$c_t - \operatorname{div}\left(D(\theta, c)\nabla c\right) = 0 \tag{3.1.2}$$

with boundary conditions

$$\frac{\partial \theta}{\partial \nu} + h(x, \theta, \theta_\Gamma(x, t)) = 0 \tag{3.1.3}$$

$$-D(\theta, c)\frac{\partial c}{\partial \nu} = b(x, t) \tag{3.1.4}$$

and initial conditions

$$\theta(x,0) = \theta^0(x) \tag{3.1.5}$$
$$c(x,0) = c^0(x). \tag{3.1.6}$$

The properties of the nonlinearities r, D, and h, as well as the hypotheses on the data b, θ_Γ, θ^0, and c^0 will be specified in the next section.

This system presents differences from the one studied in the previous chapter, mainly in the nonlinear boundary condition for θ, in the boundary condition for c and in the fact that we allow the coefficient D to depend both on θ and c. Hence, in comparison with the analysis contained in Chapter 2, we conduct here a different proof of existence of solutions and of uniqueness as well.

The first part of our analysis aims to show the existence of a generalized solution (θ, c) of the system (3.1.1)–(3.1.6), under appropriate regularity assumptions, with

$$\theta \in L^\infty(0,T;H^1(\Omega)) \cap H^1(0,T;L^2(\Omega)), \quad c \in L^2(0,T;H^1(\Omega)) \cap H^1(0,T;(H^1(\Omega))').$$

Together with the existence, employing a maximum principle and a Moser iteration, we also prove that the solution $\theta(x,t)$ is positive and uniformly bounded from above in Q_T, only assuming linear growth of $r(\theta, c)$.

Then we come to the discussion of the more delicate question of uniqueness. After the usual observation that, in order to prove a uniqueness result, it is useful to proceed with analyzing the regularity of solutions, we address explicitly the question of regularity.

To the subject of nonlinear parabolic equations of second order are dedicated many monographs, among the quite recent ones, we mention [1], [31] and [35]. Since it seemed to us that no complete result is available in literature for the case of system (3.1.1)–(3.1.6), we develop a strategy technically involved but self contained and elementary as far as possible. We are able to show that, for $N \leq 3$, θ possesses the regularity

$$\nabla \theta \in L^2(0,T;L^\infty(\Omega) \cap C^\alpha(\Omega)),$$

for some $\alpha > 0$. To this aim, we proceed in several steps. Due to the nonlinearity in the boundary condition (3.1.3), we first regularize the boundary condition with a parameter $\delta > 0$ that we will eventually let tend to zero. A fundamental tool employed in our approach to the problem are the anisotropic Sobolev spaces, which are found to be particularly adequate in the treatment of equations presenting different regularity in tangential and in normal directions, that is the case of our system with associated third-order boundary conditions.

Regarding the assumptions, we notice that no assumption about the regularity of the gradient of the solution, in particular its boundedness, was made. This technique is inspired from the estimation technique proposed in [40] for elliptic equations with linear boundary conditions. Here, however, we obtain different estimates in tangential and normal directions and it is necessary to use an embedding theorem for anisotropic spaces that can be found in a separate section, proved by the methods of Besov ([6]). The last part contains the proof of uniqueness and stability for the whole system. We prove uniqueness and continuous dependence on the data, θ_Γ, b, θ^0 and c^0. The proof is based on an L^p-variant of the Gronwall lemma. If the standard Gronwall lemma can be viewed as a result of the fact that the L^∞-norm of the function v is bounded above by its weighted L^1-norm, what we show is that an *L^p-Gronwall estimate* still holds if the L^∞-norm on the left-hand side is replaced with an L^p-norm for $p > 1$.

Remark 2. Although it is not possible to describe the whole process of case hardening through this system, still the stages without phase fractions growth may be therewith correctly modelled. In this scenario, the interaction between temperature evolution and diffusion of carbon would be reflected in the carbon diffusion coefficient $D(\theta, c)$ and in the heat source term $r(\theta, c)$. The function θ_Γ would be again the external temperature of the atmosphere in the furnace and the flux of carbon through the surface of the workpiece would be expressed by the function $b(x, t)$. In this contest we point out that a relevant issue for applications is the fact that the boundary condition for the θ, if interpreted as temperature, encompasses heat exchanges by conduction, convection and radiation. This is also a reason why we require no growth restriction on $h(x, \theta, \theta_\Gamma)$, and the boundary condition (3.1.3) thus includes thus the case

$$\frac{\partial \theta}{\partial \nu} + \alpha(x)(\theta - \theta_\Gamma) + \beta(x)(\theta^4 - \theta_\Gamma^4) = 0,$$

with coefficients $\alpha(x), \beta(x) \geq 0$, $\alpha(x) + \beta(x) \geq \alpha_0 > 0$.

3.2 Main results

We state for system (3.1.1)–(3.1.6) two sets of hypotheses: Hypothesis 3.2.1 for existence and its stronger version 3.2.2 for regularity and uniqueness. Note that we do not assume any upper bound for the growth of h.

Hypothesis 3.2.1. *The domain $\Omega \subset \mathbb{R}^N$, $N \leq 3$, is bounded and has Lipschitz-continuous boundary.*
We prescribe the data $b \in L^2(\partial\Omega \times (0,T))$, $c^0 \in L^2(\Omega)$, $\theta^0 \in V \cap L^\infty(\Omega)$, and let $0 < \theta_* := \operatorname{ess\,inf} \theta^0$.
The function h is measurable in x, locally Lipschitz-continuous with respect to θ and Lipschitz-continuous with respect to θ_Γ, with the properties

$$h(x, \theta_*, \theta_\Gamma) \leq 0,$$

$$\exists a > 0: \ \forall m > 0 \ \exists C_m > 0: \ \theta_\Gamma \leq m, \theta \geq 0 \ \Rightarrow \ h(x, \theta, \theta_\Gamma) \geq a\theta - C_m.$$

Furthermore,

- $\theta_\Gamma \in L^\infty(\partial\Omega \times (0,T))$, $(\theta_\Gamma)_t \in L^2(\partial\Omega \times (0,T))$, $\theta_\Gamma \geq \theta_*$ a. e.,
- r is Lipschitz-continuous with respect to both variables, $0 \leq r$
- D is continuous and there exist constants d_0, d_1 such that

$$0 < d_0 \leq D(\theta, c) \leq d_1$$

Hypothesis 3.2.2. *In addition to Hypothesis 3.2.1, we assume that the domain Ω is of class $C^{2,1}$, that is, the outward normal vector has Lipschitz-continuous derivatives until the second order. There exist connected relatively open subsets Γ_j of $\partial\Omega$, $j = 1, \ldots, n$, $\Gamma_i \cap \Gamma_j = \emptyset$, $i \neq j$, which are $C^{2,1}$-diffeomorphic to open bounded subsets of $\mathbb{R}^2$, and a function $h_0 \in W^{2,\infty}(\partial\Omega)$ such that $h(x, \theta, \theta_\Gamma) = h_0(x)(\theta - \theta_\Gamma)$ on $\partial\Omega \setminus \bigcup_{j=1}^n \Gamma_j$. Furthermore,*

- $\theta^0 \in W^{2,2}(\Omega)$,
- h is of class $W^{2,\infty}_{loc}$ with respect to all variables,
- $\theta_\Gamma \in L^2(0,T; W^{2,2}(\partial\Omega))$, $(\theta_\Gamma)_t \in L^2(0,T; W^{1,2}(\partial\Omega))$,
- $r, F, \partial_\theta F$ are globally Lipschitz-continuous with respect to both variables θ and c, where we set

$$F(\theta, c) = \int_0^c D(\theta, c') \,\mathrm{d}c'.$$

- $\partial_\theta F$ is globally bounded.

We deal with the following weak formulation of (3.1.1)–(3.1.4).

$$\int_\Omega (\theta_t \varphi + \nabla\theta \cdot \nabla\varphi - r(\theta,c)\varphi) \,\mathrm{d}x + \int_{\partial\Omega} (h(x,\theta,\theta_\Gamma)(x,t))\varphi \,\mathrm{d}S = 0 \quad (3.2.1)$$

$$\int_\Omega (c_t \psi + D(\theta,c)\nabla c \cdot \nabla\psi) \,\mathrm{d}x + \int_{\partial\Omega} b(x,t)\psi \,\mathrm{d}S = 0 \quad (3.2.2)$$

3.2. MAIN RESULTS

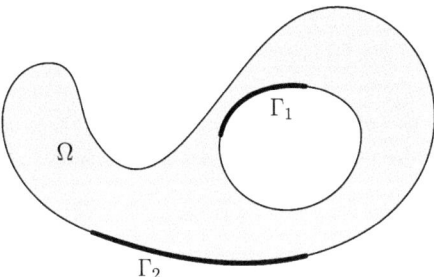

Figure 3.1: An admissible domain Ω. Thick lines denote the set of nonlinear boundary conditions.

for every test functions $\varphi, \psi \in V$.

Theorem 3.2.3. (Existence) *Let Hypothesis 3.2.1 hold. Then there exists $K_0 > 0$ and a solution (θ, c) to the system (3.2.1)–(3.2.2) with initial conditions (3.1.5)–(3.1.6), with the regularity $\theta \in L^\infty(0, T; V)$, $\theta_t \in L^2(Q_T)$, $c \in L^2(0, T; V)$, $c_t \in L^2(0, T; V')$, and such that $\theta_* \leq \theta(x, t) \leq K_0$ a. e.*

Theorem 3.2.4. (Regularity) *Let Hypothesis 3.2.2 hold. Then every solution (θ, c) to (3.2.1)–(3.2.2) from Theorem 3.2.3 has the additional regularity $\nabla \theta \in L^2(0, T; L^\infty(\Omega))$.*

To simplify the notation, we introduce the symbol

$$|w(t)|_p = \left(\int_\Omega |w(x, t)|^p \mathrm{d}x \right)^{1/p} \quad \text{for } t \in (0, T), \tag{3.2.3}$$

to denote the partial $L^p(\Omega)$-norm of a generic function $w : Q_T \to \mathbb{R}^d$, $d \geq 1$, with an obvious modification for $p = \infty$.

An important achievement of this chapter is the following uniqueness and continuous dependence result. It will be based on the partial Kirchhoff transform

$$u = F(\theta, c) \tag{3.2.4}$$

with F from Hypothesis 3.2.2.

Theorem 3.2.5. (Uniqueness and continuous data dependence) *Let Hypothesis 3.2.2 hold, and let $(\theta_1, c_1), (\theta_2, c_2)$ be two solutions with the regularity from Theorem 3.2.4 corresponding to the same $h(x, \cdot)$, and to different data $\theta_i^0, c_i^0, \theta_{\Gamma i}, b_i, i = 1, 2$, satisfying Hypothesis 3.2.2. Let $u_i = F(\theta_i, c_i), i = 1, 2$, be defined by the Kirchhoff transform (3.2.4). Set $\bar\theta = \theta_1 - \theta_2, \bar u = u_1 - u_2, \bar c^0 = c_1^0 - c_2^0, \bar\theta^0 = \theta_1^0 - \theta_2^0, \bar\theta_\Gamma = \theta_{\Gamma 1} - \theta_{\Gamma 2}, \bar b = b_1 - b_2$. Then there exists a positive constant $M > 0$ depending only on $\theta_1^0, c_1^0, \theta_{\Gamma 1}, b_1$, such that the inequality*

$$|\bar\theta(t)|_2^2 + \left|\nabla \int_0^t \bar u(\tau) d\tau\right|_2^2 + \int_0^t \left(|\nabla\bar\theta|_2^2 + |\bar u|_2^2\right)(\tau) d\tau \leq M\,\alpha(t), \qquad (3.2.5)$$

holds for every $t \in [0, T]$, where we set

$$\alpha(t) = |\bar\theta^0|_2^2 + |\bar c^0|_2^2 + \int_0^t \int_{\partial\Omega} |\bar\theta_\Gamma(x, \tau)|^2 dS\, d\tau + \int_0^t \int_{\partial\Omega} |\bar b(x, \tau)|^2 dS\, d\tau. \qquad (3.2.6)$$

3.3 Proof of existence

We fix some $K > 0$ that will be specified later, and set

$$h_K(x, \theta, \theta_\Gamma) = h(x, \max\{\theta_*, \min\{\theta, K\}\}, \theta_\Gamma).$$

Instead of (3.2.1)–(3.2.2), we consider the decoupled and truncated problem

$$\int_\Omega \left(\theta_t \varphi + \nabla\theta \cdot \nabla\varphi - r(\theta, \hat c)\,\varphi\right) dx + \int_{\partial\Omega} h_K(x, \theta, \theta_\Gamma(x, t))\,\varphi\, dS = 0 \qquad (3.3.1)$$

$$\int_\Omega \left(c_t \psi + D(\hat\theta, \hat c)\nabla c \cdot \nabla\psi\right) dx + \int_{\partial\Omega} b(x, t)\psi\, dS = 0 \qquad (3.3.2)$$

for every test functions $\varphi, \psi \in V$, with given functions $\hat\theta, \hat c \in L^2(Q_T)$, and with initial conditions (3.1.5)–(3.1.6). We now use the Schauder fixed point theorem. For $m_c, m_\theta > 0$, we fix the set

$$Z(m_\theta, m_c) = \left\{(\theta, c) \in L^2(Q_T) \times L^2(Q_T) : \int_0^T |\theta(t)|_2^2\, dt \leq m_\theta, \int_0^T |c(t)|_2^2\, dt \leq m_c\right\}.$$

Approximating the functions θ and c by Faedo-Galerkin expansions into the system of eigenfunctions of the Laplacian with homogeneous Neumann boundary conditions (see [36]), we obtain by compactness argument the existence of solutions to (3.3.1)–(3.3.2),

3.3. PROOF OF EXISTENCE

as well as the estimates

$$\int_0^T (|\theta_t(t)|_2^2 + |\Delta\theta(t)|_2^2)\,\mathrm{d}t + \sup_{t\in(0,T)} |\nabla\theta(t)|_2^2 \le R_1\left(1 + \int_0^T |\hat{c}(t)|_2^2\,\mathrm{d}t\right), \quad (3.3.3)$$

$$\int_0^T (|c_t(t)|_{V'}^2 + |\nabla c(t)|_2^2)\,\mathrm{d}t + \sup_{t\in(0,T)} |c(t)|_2^2 \le R_2, \quad (3.3.4)$$

where R_1, R_2 are constants independent of $\hat{\theta}$ and $\hat{c}$. It now suffices to choose $m_c = TR_2$ and $m_\theta = TR_1(1+TR_2)$ to check that the solution (θ, c) belongs to $Z(m_\theta, m_c)$ whenever $(\hat{\theta}, \hat{c}) \in Z(m_\theta, m_c)$. $Z(m_\theta, m_c)$ is closed, convex and the solution mapping associated with (3.3.1)–(3.3.2) is compact in $L^2(Q_T) \times L^2(Q_T)$, hence it admits a fixed point, which is a solution to (3.2.1)–(3.2.2) with h replaced by h_K.

It remains to find uniform bounds $\theta_* \le \theta \le K_0$ independent of K. Choosing $K > K_0$, we eventually obtain the assertion.

To do so, we first choose in (3.3.1) $\varphi = -(\theta_* - \theta)^+$, where z^+ denotes the positive part of an element $z \in \mathbb{R}$. We obtain

$$\frac{1}{2}\frac{\mathrm{d}}{\mathrm{d}t}\int_\Omega |(\theta_* - \theta)^+|^2\,\mathrm{d}x + \int_\Omega |\nabla(\theta_* - \theta)^+|^2\,\mathrm{d}x \le 0,$$

hence $\theta(x,t) \ge \theta_*$ a.e.

The upper bound is obtained by Moser iterations as in [33]. Set $f(x,t) = r(\theta(x,t), c(x,t))$ and $\theta_K = \min\{\theta, K\}$. Estimates (3.3.4)–(3.3.3), Sobolev embeddings, and interpolations in Lebesgue spaces yield $f \in L^2(0,T; L^6(\Omega)) \cap L^\infty(0,T; L^2(\Omega)) \subset L^q(Q_T)$ for $q = 10/3$.

The function θ is a solution of the equation

$$\int_\Omega (\theta_t\varphi + \nabla\theta\cdot\nabla\varphi - f(x,t)\varphi)\,\mathrm{d}x + \int_{\partial\Omega} h(x,\theta_K,\theta_\Gamma(x,t))\,\varphi\,\mathrm{d}S = 0 \quad (3.3.5)$$

for every $\varphi \in V$. We may choose in particular $\varphi = p\theta_K^{p-1}$ for $p > 1$, with the intention to let p tend to ∞. In the remaining part of this section, we denote by C any constant independent of K and p. Setting $v_{Kp} = \theta_K^{p/2}$, we obtain from (3.3.5) after integration with respect to t that

$$|v_{Kp}(t)|_2^2 + \int_0^t |\nabla v_{Kp}(\tau)|_2^2\,\mathrm{d}\tau + ap\int_0^t\int_{\partial\Omega} |v_{Kp}(x,\tau)|^2\,\mathrm{d}S\,\mathrm{d}\tau \quad (3.3.6)$$
$$\le C^p + Cp\left(\int_0^t\int_\Omega |f||v_{Kp}|^{2/p'}\,\mathrm{d}x\,\mathrm{d}\tau + \int_0^t\int_{\partial\Omega} |v_{Kp}(x,\tau)|^{2/p'}\,\mathrm{d}S\,\mathrm{d}\tau\right),$$

where prime denotes here and in the sequel the conjugate exponent. Using Hölder's inequality, we eliminate the boundary integrals and obtain

$$|v_{Kp}(t)|_2^2 + \int_0^t |\nabla v_{Kp}(\tau)|_2^2 \, d\tau \leq C^p + Cp \int_0^t \int_\Omega |f| \, |v_{Kp}|^{2/p'}(x,\tau) \, dx \, d\tau \quad (3.3.7)$$
$$\leq C^p + Cp\|f\|_q \|v_{Kp}\|_{2q'}^{2/p'},$$

where we use for simplicity the notation

$$\|v\|_r = \left(\int_0^T \int_\Omega |v(x,t)|^r \, dx \, dt \right)^{1/r}$$

for $v \in L^r(\Omega \times (0,T))$ and $r \geq 1$. Set $q_0 = (N/2)+1$. Then $q_0 < q$ and we define $\varrho > 0$ by the formula $q_0' = (1+\varrho)q'$. From the Gagliardo-Nirenberg inequality we obtain the estimate

$$\|v_{Kp}\|_{2q_0'}^2 \leq C \left(\sup_{t \in (0,T)} |v_{Kp}(t)|_2^2 + \int_0^T |\nabla v_{Kp}(\tau)|_2^2 \, d\tau \right),$$

hence, by virtue of (3.3.7) and Young's inequality, we obtain

$$\|v_{Kp}\|_{2q_0'}^2 \leq Cp \max\left\{ 1, C^p, \|v_{Kp}\|_{2q'}^2 \right\}, \quad (3.3.8)$$

that is,

$$\|\theta_K\|_{pq_0'} \leq (Cp)^{1/p} \max\left\{ C, \|\theta_K\|_{pq'} \right\}, \quad (3.3.9)$$

with a constant C independent of K and p. We now set $p_j = (1+\varrho)^j$, $z_j = \|\theta_K\|_{p_j q_0'}$, and $y_j = \max\{C, z_j\}$ for $j = 0, 1, 2, \ldots$. Then (3.3.9) has the form

$$y_j \leq (Cp_j)^{1/p_j} y_{j-1} \quad \text{for } j \in \mathbb{N}. \quad (3.3.10)$$

This can be rewritten as

$$\log y_j \leq C(1+\varrho)^{-j}(1+j) + \log y_{j-1} \quad \text{for } j \in \mathbb{N}, \quad (3.3.11)$$

hence the sequence y_j is bounded by a constant C independent of K. Consequently, there exists K_0 such that

$$\|\theta_K\|_p \leq K_0 \quad (3.3.12)$$

independently of p and K, which is the desired estimate that enables us to complete the proof of Theorem 3.2.3. ∎

3.4 Proof of regularity

We do not know any reference for the claim of Theorem 3.2.4, although it might be possible to obtain the same result using other strategies. The following proof is rather involved but relies essentially on elementary tools (opportune estimates in Hilbert spaces). More specifically, the proof is organized as follows.

- First of all we work, in order, with the domain: first the half space in $\mathbb{R}^N$, then the deformed half-space (as given by (3.4.3)). Eventually we will transpose the obtained results to the original bounded domain Ω of Hypothesis 3.2.2 through a localization argument (partition of unity).

- Since we know that we are going to use a localization argument to pass from the unbounded to the bounded domain, we start treating, instead of equation (3.2.1), its (slight) generalization (3.4.1), which contains a linear term more $(A\nabla v + B)$.

- As already said, we start in the case that Ω is the half-space in $\mathbb{R}^N$. Using an estimation technique which is presented similarly in the book from Nečas [40] for elliptic equations with linear boundary conditions, we prove that a solution to (3.4.1) possesses the desired regularity $\nabla v \in L^2(0,T;L^\infty(\Omega))$. This will be done in five steps: Lemma 3.4.3, 3.4.4, 3.4.5, 3.4.6 and 3.5.3.

  We point out that the nonlinear boundary condition $h(x,\theta,\theta_\Gamma)$ necessitates special attention, since the trace of v and of its first and second partial derivatives on the boundary will be involved. The main idea for treating this nonlinear term is to introduce equation (3.4.10), which contains an additional boundary term just in order to manage the trace of v at the boundary. The aim will be to let eventually tend δ to 0 and translate the results from equation (3.4.10) to equation (3.4.9).

  After this, we will dispose of the regularity of the partial derivatives of v sufficient in order to apply an opportune embedding theorem in the set of the anisotropic Sobolev spaces. This indeed requires less for the derivative in the normal direction than for the derivatives in the tangential directions. In this part we will make use of the results contained in the book from Besov, V. P. Il'in, S. M. Nikol'skiĭ [6].

- Then we consider an Ω of the form given in (3.4.3), i.e. a "deformed" half-space and also in this case we proof that the solution of 3.4.1 possesses the desired regularity $\nabla v \in L^2(0,T;L^\infty(\Omega))$. This is the content of Theorem 3.4.2.

54 CHAPTER 3. ANALYSIS OF A RELATED QUASILINEAR PARABOLIC SYSTEM

- All the estimates derived in Lemmas 3.4.3-3.4.5 (concerning the partial derivatives of v) can be transposed to the case of Ω deformed half-space. This result is contained in Theorem 3.4.1.

- The last step consists in transposing the results to the original bounded domain Ω, by means of a partition of unity argument. This leads to Theorem 3.2.4.

Let us start by considering the equation:

$$\int_\Omega (v_t \varphi + A(x)\nabla v \cdot \nabla \varphi + B(x,t) \cdot \nabla \varphi - f(x,t)\varphi)\, dx + \int_{\partial\Omega} h(x, v, v_\Gamma(x,t))\varphi\, dS = 0 \quad (3.4.1)$$

for every test function $\varphi \in W^{1,2}(\Omega)$, with initial condition $v(x,0) = v^0(x)$. Here $A = (A_{ij})_{i,j=1}^N : \Omega \to \mathbb{R}_{\text{sym}}^{N \times N}$ is a symmetric matrix function such that there exists $\kappa > 0$ with the property

$$\forall \xi \in \mathbb{R}^N : \ A(x)\xi \cdot \xi \geq \kappa |\xi|^2 \quad \text{a. e.} \quad (3.4.2)$$

and $f : Q_T \to \mathbb{R}$, $B : Q_T \to \mathbb{R}^N$, $B = (B_1, ..., B_N)$, $h : \partial\Omega \times \mathbb{R}^2 \to \mathbb{R}$, and $v_\Gamma : \partial\Omega \times (0,T) \to \mathbb{R}$ are given functions.

The reasons for introducing the functions $A(x)$ and $B(x,t)$, which do not appear in (3.2.1), are purely technical. They arise as a result of deformations of the domain and partition of unity.

Let us also specify the notations concerning the domain, which will be used in the following proofs.
Consider a set $\Omega \subset \mathbb{R}^N$ of the form

$$\Omega = \{(x', x_N) : x' \in \mathbb{R}^{N-1},\ x_N > g(x')\}, \quad (3.4.3)$$

with a given function g. Assume moreover that there exists a ball Ω_{k_0} centered at 0 of radius $k_0 > 0$ such that

$$B(x, \cdot) = f(x, \cdot) = h(x, \cdot, \cdot) = 0 \quad \text{for} \ \ x \in \Omega \setminus \Omega_{k_0}. \quad (3.4.4)$$

This assumption does not represent a restriction to our aim since we will deal with a localization argument. In particular, it is defined the space:

$$\Omega = \mathbb{R}_+^N := \{(y', y_N) : y' \in \mathbb{R}^{N-1}, y_N > 0\}. \quad (3.4.5)$$

3.4. PROOF OF REGULARITY

We will prove the following regularity results.

Theorem 3.4.1. *Let Ω be as in (3.4.3), and let $g \in W^{2,\infty}(\mathbb{R}^{N-1})$. We make the following assumptions:*

- *h is a globally Lipschitz-continuous function in all variables; furthermore, with $v, v_\Gamma \in \mathbb{R}$ fixed, the functions $h(\cdot, v, v_\Gamma), \partial_\ell h(\cdot, v, v_\Gamma)$ belong to $L^2(\partial\Omega)$ for all $\ell = 1, \ldots, N-1$;*
- *$A \in W^{1,\infty}(\Omega; \mathbb{R}^{N \times N}_{\text{sym}})$, $B \in L^2(0, T; W^{1,2}(\Omega; \mathbb{R}^N))$, $B_t \in L^2(Q_T; \mathbb{R}^N)$;*
- *$v^0 \in W^{1,2}(\Omega)$, $f \in L^2(Q_T)$, $v_\Gamma \in L^2(0, T; W^{1,2}(\partial\Omega))$, $(v_\Gamma)_t \in L^2(0, T; L^2(\partial\Omega))$.*

Let $v \in L^2(0, T; V)$ such that $v_t \in L^2(0, T; V')$ be a solution to (3.4.1). Then v has the regularity

$$v_t \in L^2(Q_T), \quad v \in L^2(0, T; W^{2,2}(\Omega)), \quad \nabla v \in L^\infty(0, T; L^2(\Omega)).$$

Theorem 3.4.2. *Let Ω be as in (3.4.3), and let $g \in W^{3,\infty}(\mathbb{R}^{N-1})$. We make the following assumptions:*

- *h is of class $W^{2,\infty}$ with respect to all variables; furthermore, with $v, v_\Gamma \in \mathbb{R}$ fixed, the functions $h(\cdot, v, v_\Gamma)$, $\partial_\ell h(\cdot, v, v_\Gamma)$, $\partial_\ell \partial_m h(\cdot, v, v_\Gamma)$ belong to $L^2(\partial\Omega)$ for all $\ell, m = 1, \ldots, N-1$;*
- *$A \in W^{2,\infty}(\Omega; \mathbb{R}^{N \times N}_{\text{sym}})$, $B \in L^2(0, T; W^{2,2}(\Omega; \mathbb{R}^N))$, $B_t \in L^2(Q_T; \mathbb{R}^N)$;*
- *$v^0 \in W^{2,2}(\Omega)$, $f \in L^2(0, T; W^{1,2}(\Omega))$, $v_\Gamma \in L^2(0, T; W^{2,2}(\partial\Omega))$, $(v_\Gamma)_t \in L^2(0, T; W^{1,2}(\partial\Omega))$.*

Let $v \in L^2(0, T; V)$ such that $v_t \in L^2(0, T; V')$ be a solution to (3.4.1). If $N \leq 3$, then v has the regularity

$$v_t \in L^2(Q_T), \quad v \in L^2(0, T; W^{2,2}(\Omega)) \quad \nabla v \in L^2(0, T; C(\bar{\Omega}).$$

Still for theorem 3.4.1 we are not aware of any reference. Therefore we proceed step by step, showing how Theorem 3.4.1 and to Theorem 3.4.2 follow from the estimates contained in the lemmas below. Before passing to the lemmas, we recall some classical estimates that are needed both here and afterwards in the proof of uniqueness.

- Young's inequality:

$$ab \leq \epsilon a^p + C_\epsilon b^{p'} \quad \forall a \geq 0, \ b \geq 0$$

- Hölder inequality:
$$\|f\|_r \leq \|f\|_p^\alpha \|f\|_q^{1-\alpha} \qquad \frac{1}{r} = \frac{\alpha}{p} + \frac{1-\alpha}{q}, \qquad p \leq r \leq q,$$
for $f \in L^p \cap L^q$, $1 \leq p \leq q \leq \infty$.

- Minkowski inequality:
$$\left(\int_Y \left(\int_X f(x,y)dx\right)^p\right)^{1/p} dy \leq \int_X \left(\int_Y f^p(x,y)dy\right)^{1/p} dx, \quad 1 < p < \infty.$$

As mentioned before, we first consider the case of Ω half-space of the form (3.4.5). For a general function $w \in W^{1,2}(\mathbb{R}^N_+)$, we have the identity
$$w^2(y', y_N) - w^2(y', 0) = 2 \int_0^{y_N} w(y', z)\, \partial_N w(y', z)\, dz,$$
hence, for every $M > 0$, integrating w.r.t. y_N, we have by Fubini Theorem that
$$w^2(y', 0) \leq \frac{1}{M} \int_0^M w^2(y', y_N)\, dy_N + 2 \int_0^M |w(y', z)|\, |\partial_N w(y', z)|\, dz.$$
Letting M tend to ∞, we obtain the trace interpolation formula
$$|w(\cdot, 0)|^2_{L^2(\mathbb{R}^{N-1})} \leq 2|w|_2\, |\partial_N w|_2 \qquad (3.4.6)$$
or, as a consequence,
$$\forall \varepsilon > 0 \ \exists C_\varepsilon > 0 \ \forall w \in W^{1,2}(\mathbb{R}^N_+) : \ |w(\cdot, 0)|^2_{L^2(\mathbb{R}^{N-1})} \leq C_\varepsilon |w|^2_2 + \varepsilon |\nabla w|^2_2. \quad (3.4.7)$$

For domains of the form (3.4.3) with a Lipschitz-continuous function g, this inequality reads after substitution in the integrals as
$$\forall \varepsilon > 0 \ \exists C_\varepsilon > 0 \ \forall w \in W^{1,2}(\Omega) : \ \int_{\partial \Omega} |w(x)|^2\, dS \leq C_\varepsilon |w|^2_2 + \varepsilon |\nabla w|^2_2. \quad (3.4.8)$$

Proceeding for instance as in [40], by a partition of unity argument, we obtain (3.4.8) for every Lipschitz-continuous domain Ω.

3.4. PROOF OF REGULARITY

If Ω is the half-space, i.e. in the context of (3.4.5), we rewrite equations (3.4.1) as

$$\int_{\mathbb{R}^N_+} (v_t\,\varphi + (A\nabla v + B)\cdot\nabla\varphi - f\,\varphi)\,\mathrm{d}y + \int_{\mathbb{R}^{N-1}} h(y', v(y',0,t), v_\Gamma(y',t))\,\varphi(y',0)\,\mathrm{d}y' = 0. \tag{3.4.9}$$

Ad mentioned before, we introduce a sort of 'regularized' problem. Let $\delta \geq 0$ and $\nabla_{y'}$ denote the partial gradient $\nabla_{y'} v = (\partial_1 v, \ldots, \partial_{N-1} v)$. We consider the equation

$$\int_{\mathbb{R}^N_+} (v_t\,\varphi + (A\nabla v + B)\cdot\nabla\varphi - f\,\varphi)\,\mathrm{d}y \tag{3.4.10}$$
$$+ \int_{\mathbb{R}^{N-1}} (\delta\,\nabla_{y'} v(y',0,t)\cdot\nabla_{y'}\varphi(y',0) + h(y', v(y',0,t), v_\Gamma(y',t))\,\varphi(y',0))\,\mathrm{d}y' = 0$$

which has to be satisfied in the case $\delta > 0$ for every test function $\varphi(y', y_N)$ from the space

$$W = \{\varphi \in W^{1,2}(\mathbb{R}^N_+) : \varphi(\cdot, 0) \in W^{1,2}(\mathbb{R}^{N-1})\}.$$

Our goal will be to derive bounds for its solution independent of δ, which then imply the corresponding estimates for the solution of (3.4.9).

Lemma 3.4.3. *Let $v^0 \in W^{1,2}(\mathbb{R}^N_+)$, $f \in L^2(\mathbb{R}^N_+ \times (0,T))$, $A \in W^{1,\infty}(\mathbb{R}^N_+; \mathbb{R}^{N\times N}_{\mathrm{sym}})$, $B \in L^2(0,T; W^{1,2}(\mathbb{R}^N_+; \mathbb{R}^N))$, and $v_\Gamma \in L^2(0,T; W^{1,2}(\mathbb{R}^{N-1}))$ be given.*
Let exist a function $h_1 \in L^2(\mathbb{R}^{N-1}) \cap L^\infty(\mathbb{R}^{N-1})$ such that h together with all its first derivatives is bounded above by h_1.
Then there exists a constant $C_1 > 0$ independent of $\delta \geq 0$ such that the solution v to (3.4.10) satisfies for all $t \in [0, T]$ the estimate

$$|\partial_\ell v(t)|_2^2 + \int_0^T |\nabla\partial_\ell v(t)|_2^2\,\mathrm{d}t + \delta \int_0^T \int_{\mathbb{R}^{N-1}} |\partial_\ell \partial_m v(y',0,t)|^2\,\mathrm{d}y'\,\mathrm{d}t \leq C_1 \tag{3.4.11}$$

for all $\ell, m = 1, \ldots, N-1$.

Proof. A solution $v \in L^2(0,T;W)$ such that $v_t \in L^2(0,T;W')$ to (3.4.10) can be constructed e.g. as follows. For $k > k_0$ (cf. (3.4.4)), we denote by $\mathbb{R}^N_k$ the intersection of $\mathbb{R}^N_+$ with the ball Ω_k, and by Γ^0_k, Γ^1_k the flat and the curved part of the boundary of $\mathbb{R}^N_k$, respectively. Instead of (3.4.10), we consider the problem

$$\int_{\mathbb{R}^N_k} (v_t\,\varphi + (A\nabla v + B)\cdot\nabla\varphi - f\,\varphi)\,\mathrm{d}y \tag{3.4.12}$$
$$+ \int_{\Gamma^0_k} (\delta\,\nabla_{y'} v(y',0,t)\cdot\nabla_{y'}\varphi(y',0) + h(y', v(y',0,t), v_\Gamma(y',t))\,\varphi(y',0))\,\mathrm{d}y' = 0,$$

with zero Dirichlet boundary condition on $\bar{\Gamma}_k^1$. We define the spaces

$$W_k = \{\varphi \in W^{1,2}(\mathbb{R}_k^N) : \varphi(\cdot, 0) \in W^{1,2}(\Gamma_k^0), \varphi = 0 \text{ on } \bar{\Gamma}_k^1\}$$

Using Galerkin approximations, the compactness lemma in [36, Section 1.5], and the compact embedding of W_k in the space $H_k = \{\varphi \in L^2(\mathbb{R}_k^N) : \varphi(\cdot, 0) \in L^2(\Gamma_k^0)\}$, we prove the existence of a solution v_k to (3.4.12), which we extend by 0 to $\mathbb{R}_+^N$. The solution is unique and satisfies the bound

$$|v_k(T)|_2^2 + \int_0^T |\nabla v_k(t)|_2^2 \, dt + \delta \int_0^T \int_{\mathbb{R}^{N-1}} |\nabla_{y'} v_k(y', 0, t)|^2 \, dy' \, dt \leq C_0 \quad (3.4.13)$$

by virtue of (3.4.7), (3.4.4), and Gronwall lemma, with a constant C_0 independent of δ and k. This, and the fact that the nonlinear term has compact support independent of k, enable us to pass to the limit as $k \to \infty$ and find a solution v to (3.4.10) satisfying the estimate (3.4.13).

To obtain higher order estimates, we use the method of the difference quotients to approximate the weak derivatives of v. In this sense it is used for example in [14] in order to improve the regularity of solutions of elliptic partial differential equations. We denote by e_ℓ for $\ell = 1, \ldots, N$ the ℓ-th unit coordinate vector, and by D_s^ℓ for $s \neq 0$ the linear mapping

$$D_s^\ell(v)(y,t) = \frac{1}{s}(v(y + se_\ell, t) - v(y, t)).$$

We will use the following formula which hold in a generic subspace U of $\mathbb{R}^N$.

$$\int_U w D_s^\ell(v) \, dy = - \int_U v D_s^{-\ell}(w) \, dy,$$

and the properties of differences quotients in relation with weak derivatives, which can be found, for instance, in the book from Evans [14, Chap. 5, Sect. 8, Th. 5].

Now let $\varphi \in W$ be given. In (3.4.10), we choose consecutively test functions $\tilde{\varphi}(y) = \varphi(y)$ and $\tilde{\varphi}(y) = \varphi(y - se_\ell)$ for some $\ell = 1, \ldots, N-1$, and subtract the two identities. This yields, after a suitable substitution, that

$$\int_{\mathbb{R}_+^N} \Big(D_s^\ell v_t \, \varphi + \big(A(y + se_\ell)\nabla(D_s^\ell v) + (D_s^\ell A)\nabla v + D_s^\ell B\big) \cdot \nabla \varphi + f(y,t) D_{-s}^\ell \varphi\Big) dy \quad (3.4.14)$$
$$+ \int_{\mathbb{R}^{N-1}} \Big(\delta \nabla_{y'} D_s^\ell v(y', 0, t) \cdot \nabla_{y'} \varphi(y', 0) + D_s^\ell \big(h(y', v(y', 0, t), v_\Gamma(y', t))\big) \varphi(y', 0)\Big) dy'.$$

3.4. PROOF OF REGULARITY

For $\varphi(y) = D_s^\ell v(y,t)$, we have in particular the estimate

$$\frac{1}{2}\frac{d}{dt}\left|D_s^\ell v(t)\right|_2^2 + \kappa \left|\nabla(D_s^\ell v(t))\right|_2^2 + \delta \int_{\mathbb{R}^{N-1}} |\nabla_{y'} D_s^\ell v(y',0,t)|^2 \, dy' \quad (3.4.15)$$
$$\leq |\nabla A|_\infty |\nabla v(t)|_2 \left|\nabla(D_s^\ell v(t))\right|_2 + (|f(t)|_2 + |\nabla B(t)|_2) \left|\nabla(D_s^\ell v(t))\right|_2$$
$$+ \int_{\mathbb{R}^{N-1}} h_1(y')(1 + |D_s^\ell v_\Gamma(y',t)| + |D_s^\ell v(y',0,t)|) |D_s^\ell v(y',0,t)| \, dy'.$$

We can pass to the limit as $s \to 0$. From (3.4.7) we have

$$|\partial v_\ell|_{\mathbb{R}^{N-1}} \leq C_\varepsilon |\partial v_\ell|_{\mathbb{R}^{N-1}} + \varepsilon |\nabla_{y'} \partial v_\ell|_{\mathbb{R}^{N-1}}$$

From this estimate, from (3.4.13) and from Gronwall lemma we obtain

$$|\partial_\ell v(t)|_2^2 + \int_0^T |\nabla \partial_\ell v(t)|_2^2 \, dt + \delta \int_0^T \int_{\mathbb{R}^{N-1}} |\nabla_{y'} \partial_\ell v(y',0,t)|^2 \, dy' \, dt \quad (3.4.16)$$
$$\leq C\left(1 + C_0 + |\nabla v^0|_2^2 + |\nabla A|_\infty^2 + \int_0^T \left(|f(t)|_2^2 + |\nabla B(t)|_2^2\right) dt + \int_0^T |\nabla_{y'} v_\Gamma(t)|_{2,\mathbb{R}^{N-1}}^2 dt\right)$$

with a constant C independent of δ, which we wanted to prove. ∎

Lemma 3.4.4. *Under the hypotheses of Lemma 3.4.3, assume in addition that:*
— $(v_\Gamma)_t \in L^2(\mathbb{R}^{N-1} \times (0,T))$ and $B_t \in L^2(\mathbb{R}_+^N \times (0,T); \mathbb{R}^N)$.
Then there exists a constant C_2 independent of $\delta \geq 0$ such that the solution v to (3.4.10) satisfies for every $t \in [0,T]$ the estimate

$$|\nabla v(t)|_2^2 + \delta \int_{\mathbb{R}^{N-1}} |\nabla_{y'} v(y',0,t)|^2 \, dy' + \int_0^T \left(|v(t)|_{W^{2,2}(\mathbb{R}_+^N)}^2 + |v_t(t)|_2^2\right) dt \leq C_2. \quad (3.4.17)$$

Proof. We discretize Eq. (3.4.10) in time, test by the time increment of v, and let the time step tend to 0. In the limit we obtain the identity

$$\int_{\mathbb{R}_+^N} \left(|v_t|^2 - f\, v_t - B_t \cdot \nabla v\right) dy + \frac{d}{dt} \int_{\mathbb{R}_+^N} \left(\frac{1}{2}A(y)\nabla v + B\right) \cdot \nabla v \, dy \quad (3.4.18)$$
$$+ \frac{d}{dt} \int_{\mathbb{R}^{N-1}} \left(\frac{\delta}{2}|\nabla_{y'} v|^2 + \hat{h}(y',v,v_\Gamma)\right) dy' = \int_{\mathbb{R}^{N-1}} \partial_{v_\Gamma} \hat{h}(y',v,v_\Gamma)(v_\Gamma)_t \, dy',$$

where

$$\hat{h}(y',v,v_\Gamma) = \int_0^v h(y',u,v_\Gamma) \, du \,.$$

We have

$$\hat{h}(y',v,v_\Gamma) \leq \frac{h_1}{2}v^2 + |v|\,(h_1|v_\Gamma| + |h(y',0,0)|)$$

and
$$|\partial_{v_\Gamma} \hat{h}(y', v, v_\Gamma)| \leq h_1 |v|.$$

Integrating equation (3.4.18) and using the estimates above, we obtain:

$$\delta \int_{\mathbb{R}^{N-1}} |\nabla_{y'} v(y', 0, t)|^2 \, dy' + \int_0^t |v_t(\tau)|_2^2 \, d\tau + |\nabla v(t)|_2^2 \quad (3.4.19)$$
$$\leq C \Big(1 + |\nabla v^0|_2^2 + \int_0^t |f(\tau)|_2^2 d\tau + \int_{\mathbb{R}^{N-1}} |v_\Gamma(y', t)|^2 dy'$$
$$+ \int_0^t \int_{\mathbb{R}^{N-1}} |(v_\Gamma)_t(y', t)|^2 dy' \, d\tau \Big)$$

with a constant C independent of δ and t as a consequence of (3.4.7) and Gronwall lemma.

At this point we are very close to the desired estimate (3.4.17). Indeed, by Lemma 3.4.3, we have $\nabla \partial_\ell v \in L^2(\mathbb{R}^N_+ \times (0, T))$ for all $\ell = 1, \ldots, N-1$.

In order to finish the proof, we must show that $\partial_N^2 v \in L^2(\mathbb{R}^n_+)$. To this aim we now choose in Eq. (3.4.10) any test function $\varphi = \varphi_0 \in L^2(0, T; W^{1,2}(\mathbb{R}^N_+))$ with a compact support in $\mathbb{R}^N_+$. We integrate by parts in all terms except for $A_{NN} \partial_N v \, \partial_N \varphi$, and obtain an identity of the form

$$\int_0^T \int_{\mathbb{R}^N_+} A_{NN}(y) \partial_N v(y, t) \, \partial_N \varphi_0(y, t) \, dy \, dt = \int_0^T \int_{\mathbb{R}^N_+} \Psi(y, t) \, \varphi_0(y, t) \, dy \, dt \quad (3.4.20)$$

with a function $\Psi \in L^2(\mathbb{R}^N_+ \times (0, T))$. Hence,

$$\partial_N \left(A_{NN}(y) \partial_N v(y, t) \right) \in L^2(\mathbb{R}^N_+ \times (0, T)).$$

Now according to the uniform ellipticity condition (3.4.2), setting $\xi = e_n = (0, \ldots, 0, 1)$ we have

$$A_{NN}(y) \geq \kappa > 0$$

for all $y \in \mathbb{R}^n_+$. Since $A_{NN} \in W^{1,\infty}(\mathbb{R}^N_+)$, we obtain that

$$\partial_N^2 v \in L^2(\mathbb{R}^n_+).$$

Therewith the proof of Lemma 3.4.4 is complete. ∎

3.4. PROOF OF REGULARITY

Lemmas 3.4.3 and 3.4.4 enable us to rewrite Eq. (3.4.10) in strong form

$$v_t - \operatorname{div}(A(y)\nabla v + B(y,t)) - f(y,t) = 0 \quad \text{a.e. in } \mathbb{R}_+^N \times (0,T) \quad (3.4.21)$$

$$\sum_{j=1}^N A_{Nj}\partial_j v + B_N - \delta \Delta_{y'} v + h(y',v,v_\Gamma(y',t)) = 0 \quad \text{a.e. in } \mathbb{R}^{N-1} \times (0,T) \quad (3.4.22)$$

where $\Delta_{y'}$ is the Laplacian with respect to y'.

Lemma 3.4.5. *Let $N \leq 3$ and $\delta > 0$. Under the hypotheses of Lemma 3.4.4, assume in addition that:*

- $v^0 \in W^{2,2}(\mathbb{R}_+^N)$, $f \in L^2(0,T;W^{1,2}(\mathbb{R}_+^N))$,

- $A \in W^{2,\infty}(\mathbb{R}_+^N;\mathbb{R}_{\text{sym}}^{N\times N})$, $B \in L^2(0,T;W^{2,2}(\mathbb{R}_+^N;\mathbb{R}^N))$,

- $v_\Gamma \in L^2(0,T;W^{2,2}(\mathbb{R}^{N-1}))$, $(v_\Gamma)_t \in L^2(0,T;W^{1,2}(\mathbb{R}^{N-1}))$,

- *there exists a function $h_2 \in L^2(\mathbb{R}^{N-1}) \cap L^\infty(\mathbb{R}^{N-1})$ such that h together with all its first and second derivatives is bounded above by h_2.*

Then there exists a constant $C_3 > 0$ independent of δ such that the solution v to (3.4.10) satisfies for all $t \in 0,T]$ the estimate

$$|\partial_m \partial_\ell v(t)|_2^2 + \int_0^T |\nabla \partial_m \partial_\ell v(t)|_2^2 \, dt + \delta \int_0^T \int_{\mathbb{R}^{N-1}} |\partial_\ell \partial_m \partial_k v(y',0,t)|^2 \, dy' \, dt \leq C_3 \quad (3.4.23)$$

for all $\ell, m, k = 1, \ldots, N-1$.

Proof. Passing to the limit in (3.4.14) as $s \to 0$, we obtain

$$\int_{\mathbb{R}_+^N} \Big(\partial_\ell v_t \, \varphi + (A\nabla \partial_\ell v + (\partial_\ell A)\nabla v + \partial_\ell B) \cdot \nabla \varphi - \partial_\ell f(y,t)\,\varphi\Big) dy \quad (3.4.24)$$
$$+ \int_{\mathbb{R}^{N-1}} (\delta \nabla_{y'} \partial_\ell v \cdot \nabla_{y'} \varphi + \partial_\ell(h(y',v(y',0,t),v_\Gamma(y',t))))\, \varphi(y',0) \, dy' = 0\,.$$

We proceed as in (3.4.14), applying to (3.4.24) the operator D_s^m with $m \in \{1,\ldots,N-1\}$, and set $\varphi(y) = D_s^m \partial_\ell v(y,t)$, with the intention to proceed as in the proof of Lemma 3.4.3. Here, the situation is more delicate because the second derivatives of the nonlinear term $h(y', v, v_\Gamma)$ will be involved. We obtain indeed the inequality

$$\begin{aligned}
\frac{1}{2}\frac{\mathrm{d}}{\mathrm{d}t}|D_s^m\partial_\ell v(t)|_2^2 &+ \kappa\,|\nabla(D_s^m\partial_\ell v(t))|_2^2 + \delta\int_{\mathbb{R}^{N-1}}|\nabla_{y'}D_s^m\partial_\ell v(y',0,t)|^2\,\mathrm{d}y' \\
&\leq \gamma(t) + C\int_{\mathbb{R}^{N-1}}(|\partial_\ell v||\partial_m v| + |\partial_\ell\partial_m v|)|\partial_\ell\partial_m v|(y',0,t)\,\mathrm{d}y' \\
&\leq \gamma(t) + \tilde{C}\int_{\mathbb{R}^{N-1}}(|\partial_\ell v|^4 + |\partial_\ell\partial_m v|^2)(y',0,t)\,\mathrm{d}y'
\end{aligned} \qquad (3.4.25)$$

where $\gamma \in L^1(0,T)$ includes all terms that have already been estimated above, and $\tilde{C}$ is a constant independent of t and δ. The right hand side of (3.4.25) is in $L^1(0,T)$ by virtue of Lemmas 3.4.3 and 3.4.4 and of the interpolation inequality:

$$|\psi|_{L^4(\mathbb{R}^{N-1})} \leq C(|\psi|_{L^2(\mathbb{R}^{N-1})} + |\psi|_{L^2(\mathbb{R}^{N-1})}^{1/2}|\nabla_{y'}\psi|_{L^2(\mathbb{R}^{N-1})}^{1/2}) \qquad (3.4.26)$$

which holds, for $N=3$, for every $\psi \in W^{1,2}(\mathbb{R}^{N-1})$ (see [8]), that we apply as follows:

$$|\partial_\ell v|_{L^4(\mathbb{R}^{N-1})} \leq C(|\partial_\ell v|_{L^2(\mathbb{R}^{N-1})} + |\partial_\ell v|_{L^2(\mathbb{R}^{N-1})}^{1/2}|\nabla_{y'}\partial_\ell v|_{L^2(\mathbb{R}^{N-1})}^{1/2}).$$

Summarizing we can see that the term

$$\tilde{C}\int_{\mathbb{R}^{N-1}}(|\partial_\ell v|^4 + |\partial_\ell\partial_m v|^2)(y',0,t)\,\mathrm{d}y'$$

is well defined but can be estimated by a constant depending on δ, since both $|\nabla_{y'}\partial_\ell v|_2$ and $|\partial_\ell\partial_m v|^2)(y',0,t)\,\mathrm{d}y'$, can be estimated in L^2, through (3.4.11), only by a constant depending on δ. This dependence will be removed later. Thanks to [8][Th.3, Chap.5, Sec.8] we can pass to the limit in (3.4.25) as $s \to 0$ and use (3.4.26) to end up with

$$\begin{aligned}
\frac{1}{2}\frac{\mathrm{d}}{\mathrm{d}t}|\partial_m\partial_\ell v(t)|_2^2 &+ \kappa\,|\nabla\partial_m\partial_\ell v(t)|_2^2 + \delta\int_{\mathbb{R}^{N-1}}|\nabla_{y'}\partial_m\partial_\ell v(y',0,t)|^2\,\mathrm{d}y' \\
&\leq \gamma(t) + C\int_{\mathbb{R}^{N-1}}(|\nabla_{y'}v|^4 + |\partial_\ell\partial_m v|^2)(y',0,t)\,\mathrm{d}y' \\
&\leq \gamma(t) + C\int_{\mathbb{R}^{N-1}}|\partial_\ell\partial_m v|^2(y',0,t)\,\mathrm{d}y' \\
&+ C\Big(|\nabla_{y'}v(\cdot,0,t)|_{L^2(\mathbb{R}^{N-1})}^4 + |\nabla_{y'}v(\cdot,0,t)|_{L^2(\mathbb{R}^{N-1})}^2|\Delta_{y'}v(\cdot,0,t)|_{L^2(\mathbb{R}^{N-1})}^2\Big)
\end{aligned} \qquad (3.4.27)$$

with a possibly different function $\gamma \in L^1(0,T)$ and different constants C independent of t and δ. To continue the estimate of the right hand side of (3.4.27), we need formula (3.4.6), that we recall here:

$$|w(\cdot,0)|_{L^2(\mathbb{R}^{N-1})}^2 \leq 2|w|_2\,|\partial_N w|_2$$

3.4. PROOF OF REGULARITY

With this we obtain the following upper bound:

$$\gamma(t) + C\Big(|\partial_\ell\partial_m v|_2 |\partial_N\partial_\ell\partial_m v|_2 + |\nabla_{y'}v|_2^2 |\partial_N \nabla_{y'}v|_2^2 + |\nabla_{y'}v|_2 |\partial_N \nabla_{y'}v|_2 |\Delta_{y'}v|_2 |\partial_N \Delta_{y'}v|_2\Big).$$

Now we observe that, by Lemma 3.4.3, $|\nabla_{y'}v|_2 \le C_1$, and $\beta(t) := |\partial_N \nabla_{y'}v(t)|_2$ belongs to $L^2(0,T)$. Hence, by Young's inequality, we obtain from (3.4.27) the estimate

$$\frac{d}{dt} \sum_{\ell,m=1}^{N-1} |\partial_m\partial_\ell v(t)|_2^2 + \sum_{\ell,m=1}^{N-1} |\nabla\partial_m\partial_\ell v(t)|_2^2$$

$$\le \gamma(t) + C \sum_{\ell,m=1}^{2} \Big(|\partial_\ell\partial_m v|_2^2 + \beta^2(t)|\Delta_{y'}v|_2^2\Big), \quad (3.4.28)$$

and from Gronwall's argument we obtain (3.4.23). ∎

We now let δ tend to 0 and prove the following step.

Lemma 3.4.6. *Under the hypotheses of Lemma 3.4.5, there exists a constant $C_4 > 0$ such that the solution v to (3.4.9) satisfies for all $t \in [0,T]$ the estimate*

$$|\partial_N\partial_\ell v(t)|_2^2 + \int_0^T |\nabla\partial_N\partial_\ell v(t)|_2^2 \, dt \le C_4. \quad (3.4.29)$$

for all $\ell = 1, \ldots, N-1$.

Proof. From Lemma 3.4.5 it follows that the solution v to (3.4.9) satisfies (3.4.21)–(3.4.23) with $\delta = 0$. Let us consider now test functions φ with compact support in $\mathbb{R}^N_+$, and apply the operator D_s^N to Eq. (3.4.9). As a counterpart of (3.4.14), we obtain

$$\int_{\mathbb{R}^N_+} \Big(D_s^N v_t \, \varphi + (A(y+se_N)\nabla(D_s^N v) + (D_s^N A)\nabla v + D_s^N R) \cdot \nabla\varphi - D_s^N f(y,t)\,\varphi\Big) dy = 0. \quad (3.4.30)$$

Passing to the limit as $s \to 0$ yields

$$\int_{\mathbb{R}^N_+} \Big(\partial_N v_t\, \varphi + \big(A(y)\nabla\partial_N v + (\partial_N A(y))\nabla v + \partial_N B(y,t)\big) \cdot \nabla\varphi - \partial_N f(y,t)\,\varphi\Big) dy = 0. \quad (3.4.31)$$

Let V_0 denote the space $W_0^{1,2}(\mathbb{R}^N_+)$. We choose any $\varphi_0 \in V_0$ and set $\varphi = A_{N\ell}\varphi_0$ in (3.4.24) and obtain

$$\int_{\mathbb{R}^N_+} \Big(\partial_\ell v_t\, A_{N\ell}\varphi_0 + \big(A\nabla\partial_\ell v + (\partial_\ell A)\nabla v + \partial_\ell B\big) \cdot \nabla A_{N\ell}\varphi_0$$
$$- \partial_\ell f(y,t)\, A_{N\ell}\varphi_0\Big) dy = 0. \quad (3.4.32)$$

Analogously we choose $\varphi = A_{NN}\,\varphi_0$ and set it in (3.4.31)

$$\int_{\mathbb{R}^N_+} \Big(\partial_N v_t\, A_{NN}\,\varphi_0 + \big(A(y)\nabla\partial_N v + (\partial_N A(y))\nabla v + \partial_N B(y,t)\big)\cdot \nabla A_{NN}\,\varphi_0 \\ - \partial_N f(y,t)\,A_{NN}\,\varphi_0\Big)dy = 0\,. \qquad (3.4.33)$$

To proceed with the computations, we need the formula

$$A\nabla \partial_\ell v \cdot \nabla(A_{N\ell}\,\varphi_0) = A\nabla(A_{N\ell}\,\partial_\ell v)\cdot \nabla \varphi_0 + (A\nabla \partial_\ell v \cdot \nabla A_{N\ell})\,\varphi_0 - \partial_\ell v(A\nabla A_{N\ell}\cdot \nabla \varphi_0)\,.$$

Now, with a computation, we can rewrite the equations (3.4.32),(3.4.33) as

$$\int_{\mathbb{R}^N_+} \Big(A_{N\ell}\,\partial_\ell v_t\,\varphi_0 + A\nabla(A_{N\ell}\partial_\ell v)\cdot \nabla\varphi_0 + \underbrace{(A\nabla\partial_\ell v\cdot \nabla A_{N\ell})\,\varphi_0}_{I} - \underbrace{\partial_\ell v(A\nabla A_{N\ell}\cdot \nabla\varphi_0)}_{II} \\ + \underbrace{(\partial_\ell A\nabla v + \partial_\ell B)\cdot \nabla(A_{N\ell}\,\varphi_0)}_{III} - A_{N\ell}\,\partial_\ell f(y,t)\,\varphi_0 \Big)dy = 0 \qquad (3.4.34)$$

for all $\ell = 1,\ldots, N$.

Consider now the function

$$w = \sum_{\ell=1}^N A_{N\ell}\partial_\ell v.$$

Summing up the above identities over ℓ and as a consequence of (3.4.22) with $\delta = 0$, we can see that w is a solution of the inhomogeneous Dirichlet problem

$$\int_{\mathbb{R}^N_+}\Big(w_t\,\varphi_0 + A(y)\nabla w\cdot \nabla\varphi_0 - f_1(y,t)\,\varphi_0\Big)dy = 0 \quad \forall \varphi_0 \in V_0\,, \qquad (3.4.35)$$

with boundary condition

$$w(y',0,t) + B_N(y',0,t) + h\Big(y',v(y',0,t),v_\Gamma(y',t)\Big) = 0 \qquad (3.4.36)$$

on $\mathbb{R}^{N-1}\times (0,T)$. The function f_1 in (3.4.35) is given by:

$$f_1 = \sum_{\ell=1}^N \Big(A_{N\ell}\partial_\ell f - \underbrace{A\nabla\partial_\ell v\cdot \nabla A_{N\ell}}_{I} - \underbrace{\mathrm{div}\,((\partial_\ell v)A\nabla A_{N\ell})}_{II} + \underbrace{A_{N\ell}\mathrm{div}\,((\partial_\ell A)\nabla v + \partial_\ell B)}_{III}\Big), \qquad (3.4.37)$$

Hence f_1 belongs to $L^2(\mathbb{R}^N_+ \times (0,T))$. We now fix a smooth function ϱ with compact support in $\mathbb{R}_+$ and such that $\varrho(0) = 1$, and set

$$w_1(y,t) = B_N(y,t) + \varrho(y_N)\,h(y',v(y,t),v_\Gamma(y',t))\,. \qquad (3.4.38)$$

3.4. PROOF OF REGULARITY

w_1 so defined can be seen as a prolongation of the boundary condition to $\mathbb{R}_+^N \times (0,T)$. The function $w_0 := w - w_1$ is a solution to the homogeneous Dirichlet problem for the following counterpart of (3.4.35)

$$\int_{\mathbb{R}_+^N} \Big((w_0)_t\,\varphi_0 + A(y)\nabla w_0 \cdot \nabla\varphi_0 - f_2(y,t)\,\varphi_0\Big)\mathrm{d}y = 0 \quad \forall \varphi_0 \in V_0, \qquad (3.4.39)$$

where

$$f_2 = f_1 - (w_1)_t + \operatorname{div} A(y)\nabla w_1. \qquad (3.4.40)$$

Let us check that $f_2 \in L^2(\mathbb{R}_+^N \times (0,T))$. By virtue of Lemmas 3.4.4–3.4.5, the delicate term in f_2 is $\operatorname{div} A(y)\nabla w_1$, since it involves, due to the nonlinearity of h terms of the form $\partial_\ell^2 v$. These terms will be bounded in L^2 provided we prove that

$$\nabla v \in L^4(\mathbb{R}_+^N \times (0,T)). \qquad (3.4.41)$$

To this end, we refer to [6, Vol I, Theorem 10.2], see also Remark 3, which states that there exists a constant $C > 0$ such that for every function $\xi \in L^2(\mathbb{R}_+^N \times (0,T))$ with the regularity $\partial_\ell^2 \xi, \partial_N \xi \in L^2(\mathbb{R}_+^N \times (0,T))$ for all $\ell = 1, \ldots, N-1$, and for every $\sigma \in (0,1]$ we have the inequality (note that $N \leq 3$!)

$$|\xi|_4 \leq C\left(\sigma^{-1/2}|\xi|_2 + \sigma^{1/2}\left(|\partial_N \xi|_2 + \sum_{\ell=1}^{N-1} |\partial_\ell^2 \xi|_2\right)\right). \qquad (3.4.42)$$

This can be equivalently written as

$$|\xi|_4 \leq C\left(|\xi|_2 + |\xi|_2^{1/2}\left(|\partial_N \xi|_2 + \sum_{\ell=1}^{N-1} |\partial_\ell^2 \xi|_2\right)^{1/2}\right). \qquad (3.4.43)$$

In (3.4.43), we choose $\xi = \partial_k v(t)$ for $k = 1, \ldots, N$ and a. e. t. From Lemmas 3.4.4–3.4.5 we obtain (3.4.41), hence $f_2 \in L^2(\mathbb{R}_+^N \times (0,T))$.

Following the same idea as in the proof of Lemma 3.4.3, we now apply the operator D_s^ℓ to Eq. (3.4.39) for $\ell = 1, \ldots, N-1$ and test by $\varphi_0 = D_s^\ell w_0$.
Using the identity $\int (D_s^\ell w_1)_t D_s^\ell w_0 \,\mathrm{d}y = \int (w_1)_t D_{-s}^\ell D_s^\ell w_0 \,\mathrm{d}y$, we may let s tend to 0 and conclude that

$$\partial_\ell \nabla w_0 \in L^2(\mathbb{R}_+^N \times (0,T))$$

for all $\ell = 1, \ldots, N-1$. By Lemma 3.4.5, and since $A_{NN} \geq \kappa$, we obtain that

$$\partial_\ell \partial_N^2 v \in L^2(\mathbb{R}_+^N \times (0,T))$$

for all $\ell = 1, \ldots, N-1$, and the proof is complete. ∎

3.5 An anisotropic embedding theorem

Here we prove here an embedding theorem for anisotropic Sobolev spaces that is needed for the proof of Theorem 3.4.1. For a vector $\mathbf{p} = (p_1, \ldots, p_N)$, $1 \leq p_i < \infty$, we define the space $L^{\mathbf{p}}(\mathbb{R}^N)$ as the subspace of $L^1(\mathbb{R}^N)$ of functions u such that the norm

$$\|u\|_{\mathbf{p}} = \left(\int_{\mathbb{R}} \left(\cdots \int_{\mathbb{R}} \left(\int_{\mathbb{R}} |u(x)|^{p_1} \, dx_1 \right)^{p_2/p_1} dx_2 \cdots \right)^{p_N/p_{N-1}} dx_N \right)^{1/p_N} \tag{3.5.1}$$

is finite. For a matrix $\mathbf{P} = (P_{ij})_{i,j=1}^N$, $P_{ij} = 1/p_{ij}$, $1 \leq p_{ij} < \infty$, we define the anisotropic Sobolev space

$$W^{1,\mathbf{P}}(\mathbb{R}^N) = \left\{ u \in L^1(\mathbb{R}^N) : \frac{\partial u}{\partial x_i} \in L^{\mathbf{p}_i}(\mathbb{R}^N), i = 1, \ldots, N \right\}, \tag{3.5.2}$$

where $\mathbf{p}_i = (p_{i1}, \ldots, p_{iN})$.

We denote by $\mathbf{I}$ the identity $N \times N$ matrix, and by $\mathbf{1}$ the vector $\mathbf{1} = (1, 1, \ldots, 1)$. The spectral radius $\varrho(\mathbf{P})$ of $\mathbf{P}$ is defined as

$$\varrho(\mathbf{P}) = \max\{|\lambda| : \lambda \in \mathbb{C}, \det(\mathbf{P} - \lambda \mathbf{I}) = 0\} = \limsup_{n \to \infty} |\mathbf{P}^n|^{1/n}. \tag{3.5.3}$$

Theorem 3.5.1. *Let $\varrho(\mathbf{P}) < 1$, and let*

$$(\mathbf{I} - \mathbf{P})^{-1} \mathbf{1} = \mathbf{b} = (b_1, \ldots, b_N). \tag{3.5.4}$$

Then $W^{1,\mathbf{P}}(\mathbb{R}^N)$ is embedded in $L^\infty(\mathbb{R}^N)$, and there exists a constant $C > 0$ such that each $u \in W^{1,\mathbf{P}}(\mathbb{R}^N)$ has for all $x, z \in \mathbb{R}^N$ the Hölder property

$$|u(z) - u(x)| \leq C \|u\|_{W^{1,\mathbf{P}}(\mathbb{R}^N)} \sum_{i=1}^N |z_i - x_i|^{1/b_i}. \tag{3.5.5}$$

The identity (3.5.4) can be written as

$$\mathbf{b} = (\mathbf{I} + \mathbf{P} + \mathbf{P}^2 + \ldots) \mathbf{1}.$$

Since all entries of $\mathbf{P}$ are positive, we obtain $b_i > 1$ for all i, so that the right hand side of

3.5. AN ANISOTROPIC EMBEDDING THEOREM

(3.5.5) is meaningful. Note also that in the isotropic case $p_{ij} = p$, Theorem 3.5.1 gives the well-known embedding condition $p > N$ with Hölder exponent $1/b = 1 - (N/p)$.

Proof. Following [6], we fix a smooth function Φ with compact support in $\mathbb{R}^N$ such that $\int_{\mathbb{R}^N} \Phi(x)\,\mathrm{d}x = 1$, and for $\sigma > 0$ and $u \in W^{1,\mathbf{P}}(\mathbb{R}^N)$ set

$$u^\sigma(x) = \sigma^{-|\mathbf{b}|} \int_{\mathbb{R}^N} \Phi\!\left(\frac{x-y}{\sigma^{\mathbf{b}}}\right) u(y)\,\mathrm{d}y\,, \tag{3.5.6}$$

where $|\mathbf{b}| = \sum_{i=1}^N b_i$ and

$$\frac{x-y}{\sigma^{\mathbf{b}}} = \left(\frac{x_1 - y_1}{\sigma^{b_1}}, \ldots, \frac{x_N - y_N}{\sigma^{b_N}}\right).$$

By substitution, we have the identity

$$u^\sigma(x) = \int_{\mathbb{R}^N} \Phi(z) u(x - \sigma^{\mathbf{b}} z)\,\mathrm{d}z\,, \tag{3.5.7}$$

which implies that

$$\lim_{\sigma \to 0} |u^\sigma - u|_1 = 0\,. \tag{3.5.8}$$

We differentiate u^σ with respect to σ, integrate by parts with respect to y, and obtain

$$\frac{\partial u^\sigma(x)}{\partial \sigma} = -\sum_{i=1}^N \sigma^{-|\mathbf{b}|-1+b_i} \int_{\mathbb{R}^N} \Psi_i\!\left(\frac{x-y}{\sigma^{\mathbf{b}}}\right) \frac{\partial u(y)}{\partial y_i}\,\mathrm{d}y\,, \tag{3.5.9}$$

where

$$\Psi_i(z) = b_i z_i \Phi(z) \quad \text{for } z \in \mathbb{R}^N.$$

By the anisotropic Hölder inequality we have

$$\left|\frac{\partial u^\sigma(x)}{\partial \sigma}\right| \leq \sum_{i=1}^N \sigma^{-|\mathbf{b}|-1+b_i} \left\|\Psi_i\!\left(\frac{\cdot}{\sigma^{\mathbf{b}}}\right)\right\|_{\mathbf{p}'_i} \left\|\frac{\partial u}{\partial y_i}\right\|_{\mathbf{p}_i}, \tag{3.5.10}$$

where $\mathbf{p}'_i$ is the componentwise conjugate of $\mathbf{p}_i$. By substitution, we have

$$\left\|\Psi_i\!\left(\frac{\cdot}{\sigma^{\mathbf{b}}}\right)\right\|_{\mathbf{p}'_i} = \sigma^{\sum_{j=1}^N b_j/p'_{ij}} \|\Psi_i\|_{\mathbf{p}'_i} = \sigma^{|\mathbf{b}|-(\mathbf{Pb})_i} \|\Psi_i\|_{\mathbf{p}'_i}. \tag{3.5.11}$$

This and (3.5.4) yield the following estimate independent of σ and x:

$$\left|\frac{\partial u^\sigma(x)}{\partial \sigma}\right| \leq \sum_{i=1}^N \sigma^{b_i - (\mathbf{Pb})_i - 1} \|\Psi_i\|_{\mathbf{p}'_i} \left\|\frac{\partial u}{\partial y_i}\right\|_{\mathbf{p}_i} = \sum_{i=1}^N \|\Psi_i\|_{\mathbf{p}'_i} \left\|\frac{\partial u}{\partial y_i}\right\|_{\mathbf{p}_i} =: U\,. \tag{3.5.12}$$

For $\sigma > \tilde{\sigma} > 0$ we have
$$|u^\sigma(x) - u^{\tilde{\sigma}}(x)| \leq (\sigma - \tilde{\sigma})U,$$
hence u^σ converges uniformly in $L^\infty(\mathbb{R}^N)$ as $\sigma \to 0$. In view of (3.5.8), its limit is u, which thus belongs to $L^\infty(\mathbb{R}^N) \cap C(\mathbb{R}^N)$, and we have for all $\sigma > 0$ the embedding inequality
$$|u(x)| \leq |u^\sigma(x)| + \sigma U \leq |\Phi|_\infty \sigma^{-|\mathbf{b}|}|u|_1 + \sigma U. \tag{3.5.13}$$
To prove the Hölder estimate, we replace $u(x)$ in (3.5.13) by $u(x + he_i) - u(x)$, where e_i is the i-th unit coordinate vector and $h > 0$ is arbitrary. We obtain
$$|u(x + he_i) - u(x)| \leq |u^\sigma(x + he_i) - u^\sigma(x)| + 2\sigma U, \tag{3.5.14}$$
where
$$\begin{aligned}u^\sigma(x + he_i) - u^\sigma(x) &= \sigma^{-|\mathbf{b}|} \int_{\mathbb{R}^N} \Phi\left(\frac{x-y}{\sigma^\mathbf{b}}\right)(u(y + he_i) - u(y))\, dy \tag{3.5.15}\\ &= -\sigma^{-|\mathbf{b}|} \int_0^h \int_{\mathbb{R}^N} \Phi\left(\frac{x-y}{\sigma^\mathbf{b}}\right)\frac{\partial u}{\partial y_i}(y + se_j)\, dy\, ds.\end{aligned}$$

This and (3.5.11) entail
$$|u^\sigma(x + he_i) - u^\sigma(x)| \leq h\sigma^{-|\mathbf{b}|}\left\|\Psi_i\left(\frac{\cdot}{\sigma^\mathbf{b}}\right)\right\|_{\mathbf{p}_i'}\left\|\frac{\partial u}{\partial y_i}\right\|_{\mathbf{p}_i} \leq h\sigma^{-(\mathbf{P}\mathbf{b})_i}\|\Psi_i\|_{\mathbf{p}_i'}\left\|\frac{\partial u}{\partial y_i}\right\|_{\mathbf{p}_i}. \tag{3.5.16}$$

We thus conclude from (3.5.14) that there exists a constant $C > 0$ such that for all $u \in W^{1,\mathbf{P}}(\mathbb{R}^N)$, $x \in \mathbb{R}^N$, $\sigma > 0$, and $h > 0$ we have
$$|u(x + he_i) - u(x)| \leq C\left(h\sigma^{-(\mathbf{P}\mathbf{b})_i} + \sigma\right)\sum_{j=1}^N \left\|\frac{\partial u}{\partial y_j}\right\|_{\mathbf{p}_j}. \tag{3.5.17}$$

In particular, for $\sigma = h^{1/b_i}$ we obtain, by virtue of (3.5.4), the formula
$$|u(x + he_i) - u(x)| \leq C h^{1/b_i} \|u\|_{W^{1,\mathbf{P}}(\mathbb{R}^N)}, \tag{3.5.18}$$
and (3.5.5) follows from the triangle inequality. ∎

Corollary 3.5.2. *The space:*
$$Y := \left\{w \in L^1(\mathbb{R}_+^N) : \partial_N w \in X^{p_0, q_0},\ \nabla_{y'} w \in X^{p_1, q_1}\right\} \tag{3.5.19}$$

3.5. AN ANISOTROPIC EMBEDDING THEOREM

satisfies the condition in Theorem 3.5.1 if and only if

$$\frac{p'_0}{p_1 q_0} + \frac{1}{q_1} < \frac{1}{N-1}. \tag{3.5.20}$$

Proof. The matrix $\mathbf{P} - \lambda \mathbf{I}$ has the form

$$\mathbf{P} - \lambda \mathbf{I} = \begin{pmatrix} 1/q_1 - \lambda & 1/q_1 & \cdots & 1/q_1 & 1/p_1 \\ 1/q_1 & 1/q_1 - \lambda & \cdots & 1/q_1 & 1/p_1 \\ & & \cdots & & \\ 1/q_1 & 1/q_1 & \cdots & 1/q_1 - \lambda & 1/p_1 \\ 1/q_0 & 1/q_0 & \cdots & 1/q_0 & 1/p_0 - \lambda \end{pmatrix},$$

and its determinant is

$$\det(\mathbf{P} - \lambda \mathbf{I}) = (-\lambda)^{N-2} \left(\left(\frac{N-1}{q_1} - \lambda \right) \left(\frac{1}{p_0} - \lambda \right) - \frac{N-1}{q_0 p_1} \right).$$

We easily check that all roots of the equation $\det(\mathbf{P} - \lambda \mathbf{I}) = 0$ are in absolute value smaller that 1 if and only if condition (3.5.20) holds. ∎

Remark 3. The embedding formula (3.4.42) in $\mathbb{R}^3$ can be derived from (3.5.6), where we set $b_1 = b_2 = \frac{1}{2}$, $b_3 = 1$. Put $u(y', y_N) = \xi(y', y_N)$ for $y_N > 0$, $u(y', y_N) = \xi(y', -y_N)$ for $y_N < 0$. Assuming that $\Phi(z) = \Phi(-z)$, we may set $\hat{\Psi}_1(z) = \int_{-\infty}^{z_1} \Psi_1(s, z_2, z_3) ds$, $\hat{\Psi}_2(z) = \int_{-\infty}^{z_2} \Psi_2(z_1, s, z_3) ds$. Then $\hat{\Psi}_1$, $\hat{\Psi}_2$ have compact support and we may integrate by parts in (3.5.9) to obtain

$$\frac{\partial u^\sigma}{\partial \sigma}(x) = -\sum_{i=1}^{2} \sigma^{-|\mathbf{b}|-1+2b_i} \int_{\mathbb{R}^3} \hat{\Psi}_i\left(\frac{x-y}{\sigma^{\mathbf{b}}}\right) \frac{\partial^2 u(y)}{\partial y_i^2} dy$$
$$- \sigma^{-|\mathbf{b}|-1+b_3} \int_{\mathbb{R}^3} \Psi_3\left(\frac{x-y}{\sigma^{\mathbf{b}}}\right) \frac{\partial u(y)}{\partial y_3} dy.$$

Integrals of the form $\int_{\mathbb{R}^3} \Psi_*\left(\frac{x-y}{\sigma^{\mathbf{b}}}\right) u_*(y) \, dy$ with $u_* \in L^2(\mathbb{R}^3)$ can be estimated in $L^4(\mathbb{R}^3)$ using the Young inequality for convolutions as

$$\left| \int_{\mathbb{R}^3} \Psi_*\left(\frac{\cdot - y}{\sigma^{\mathbf{b}}}\right) u_*(y) \, dy \right|_4 \le \sigma^{(3/4)|\mathbf{b}|} |\Psi_*|_{4/3} |u_*|_2. \tag{3.5.21}$$

Hence, by virtue of the choice of $\mathbf{b}$, we have

$$\left|\frac{\partial u^\sigma}{\partial \sigma}\right|_4 \le C\sigma^{-1/2}\left(\left|\frac{\partial^2 u}{\partial y_1^2}\right|_2 + \left|\frac{\partial^2 u}{\partial y_2^2}\right|_2 + \left|\frac{\partial u}{\partial y_3}\right|_2\right), \qquad |u^\sigma|_4 \le C\sigma^{-1/2}|u|_2, \tag{3.5.22}$$

and (4.37) follows from the inequality

$$|u|_4 \leq |u^\sigma|_4 + \left| \int_0^\sigma \frac{\partial u^{\sigma'}}{\partial \sigma'} d\sigma' \right|_4. \tag{3.5.23}$$

Proof of Theorems 3.4.1 and 3.4.2.

Now, we can conclude the proof of the regularity of the solution θ, proceeding first with the proof of Theorem 3.4.1.

We now define in particular the anisotropic spaces

$$X^{p,q} = \left\{ w \in L^1(\mathbb{R}^N_+) : \int_0^\infty \left(\int_{\mathbb{R}^{N-1}} |w(y', y_N)|^q dy' \right)^{p/q} dy_N < \infty \right\}.$$

We can extend the functions defined on $\mathbb{R}^N_+$ by symmetry to $\mathbb{R}^N$, and use Corollary 3.5.2 of the previous section to obtain the embedding

$$Y := \left\{ w \in L^1(\mathbb{R}^N_+) : \partial_N w \in X^{p_0, q_0}, \ \nabla_{y'} w \in X^{p_1, q_1} \right\} \tag{3.5.24}$$

in the space $C^\alpha(\mathbb{R}^N_+) \cap L^\infty(\mathbb{R}^N_+)$ of bounded α-Hölder continuous functions for some $\alpha > 0$, provided

$$\frac{p'_0}{p_1 q_0} + \frac{1}{q_1} < \frac{1}{N-1},$$

where p'_0 is the conjugate exponent to p_0. As a direct consequence, we have

Lemma 3.5.3. *Under the conditions of Lemma 3.4.5, we have $\nabla v \in L^2(0, T; L^\infty(\mathbb{R}^N_+))$.*

Proof. The functions $\partial_\ell v$ for $\ell = 1, \ldots, N-1$ belong to $L^2(0, T; W^{2,2}(\mathbb{R}^N_+))$, which is embedded into $L^2(0, T; L^\infty(\mathbb{R}^N_+))$ by classical Sobolev embedding theorems, see [8]. For $w(y, t) = \partial_N v(y, t)$ and a. e. $t \in (0, T)$, we have

$$|\partial_\ell w(t)|_{X^{6,6}} = |\partial_\ell w(t)|_6 \leq C \left(|\partial_\ell w(t)|_2 + |\nabla \partial_\ell w(t)|_2 \right) \quad \text{for } \ell = 1, \ldots, N-1,$$
$$|\partial_N w(t)|_{X^{2,q}} \leq C \left(|\partial_N w(t)|_2 + |\nabla_{y'} \partial_N w(t)|_2 \right)$$

with a constant $C > 0$ since for $N = 2$, $W^{1,2}(\Omega) \subset L^q(\Omega)$, for every $q \geq 2$. Hence, (3.5.24) is fulfilled with $p_0 = 2$, $q_0 = q$, $p_1 = q_1 = 6$. Integrating over t we conclude the proof of Lemma 3.5.3. ∎

This enables us to prove Theorems 3.4.1 and 3.4.2.

3.5. AN ANISOTROPIC EMBEDDING THEOREM

Proof of Theorems 3.4.1 and 3.4.2. We substitute in (3.4.1) new variables $y' = x'$, $y_N = x_N - g(x')$, and obtain for the new unknown function $\tilde{v}(y', y_N) = v(y', y_N + g(y'))$ the equation

$$\int_{\mathbb{R}^N_+} \left(\tilde{v}_t \varphi + (\tilde{A}\nabla \tilde{v} + \tilde{B}) \cdot \nabla \varphi - \tilde{f}\varphi \right) dy + \int_{\mathbb{R}^{N-1}} \tilde{h}(y', \tilde{v}(y', 0, t), v_\Gamma(y', t)) \varphi(y', 0)\, dy' = 0 \quad (3.5.25)$$

for every $\varphi \in W^{1,2}(\mathbb{R}^N_+)$, where

$$\begin{aligned}
\tilde{f}(y', y_N, t) &= f(y', y_N + g(y'), t), \\
\tilde{v}_\Gamma(y', t) &= v_\Gamma(y', g(y'), t), \\
\tilde{h}(y', v, v_\Gamma) &= h(y', g(y'), v, v_\Gamma)\sqrt{1 + |\nabla_{y'} g(y')|^2}, \\
\tilde{A}(y', y_N) &= L^T(y')\, A(y', y_N + g(y'))\, L(y'), \\
\tilde{B}(y', y_N, t) &= L^T(y')\, B(y', y_N + g(y'), t),
\end{aligned}$$

and where the matrix L has the same form as in

$$L = \begin{pmatrix} 1 & 0 & \ldots & 0 & -\partial_1 g \\ 0 & 1 & \ldots & 0 & -\partial_2 g \\ & & \ldots & & \\ 0 & 0 & \ldots & 1 & -\partial_{N-1} g \\ 0 & 0 & \ldots & 0 & 1 \end{pmatrix}.$$

Theorem 3.4.1 now follows from Lemma 3.4.4. Theorem 3.4.2 is a consequence of Lemma 3.5.3. ∎

We are now ready to prove Theorem 3.2.4.

Proof of Theorem 3.2.4. The nonlinear boundary condition is active only on the subsets Γ_j of $\partial\Omega$ for $j = 1, \ldots, n$.
We choose a covering of Ω, $\{\Omega_j\}_{j=1}^n$, $\bar{\Omega} \subset \bigcup_{j=1}^n \Omega_j$ with the property that

$$\Gamma_j \subset \Omega_j, \quad \Gamma_i \cap \bar{\Omega}_j = \emptyset \quad \text{for } i \neq j.$$

We now find a smooth partition of unity $\{\lambda_j\}_{j=1}^n$, $\lambda_j \in C^\infty(\mathbb{R}^N)$ so that

$$1 = \sum_{j=1}^n \lambda_j(x) \quad \text{on } \bar{\Omega} \quad \text{such that} \quad \operatorname{supp}\lambda_j \subset \bar{\Omega}_j,$$

(such a partition of unity exists, see for instance [8]). We set $v_j = \theta\, \lambda_j$, $f(x, t) =$

$r(\theta(x,t), c(x,t))$. After suitable deformations and rotations, we may assume that each set $\Omega_j \cap \bar{\Omega}$ can be extended to a domain $\tilde{\Omega}_j$ of the form (3.4.3), such that $\Gamma_j \subset \partial \tilde{\Omega}_j$. To derive the equation for v_j, we test the equation

$$\int_\Omega (\theta_t \varphi + \nabla\theta \cdot \nabla\varphi - f(x,t)\varphi)\,dx + \int_{\partial\Omega} h(x,\theta,\theta_\Gamma(x,t))\,\varphi\,dS = 0 \qquad (3.5.26)$$

by $\varphi = \lambda_j \tilde{\varphi}$, and obtain

$$\int_{\tilde{\Omega}_j} ((v_j)_t\,\tilde{\varphi} + \nabla v_j \cdot \nabla\tilde{\varphi} + B_j \cdot \nabla\tilde{\varphi} - f_j(x,t)\,\tilde{\varphi})\,dx + \int_{\partial\tilde{\Omega}_j} h(x, v_j, v_{\Gamma j}(x,t))\,\tilde{\varphi}\,dS = 0,$$
$$(3.5.27)$$

with $B_j = -\theta \nabla \lambda_j$, $f_j = f\lambda_j - \nabla\theta \cdot \nabla\lambda_j$, $v_{\Gamma j} = \theta_\Gamma \lambda_j$. Here we have used the fact that $\lambda_j = 1$ on Γ_j, and that h is linear on $\partial\tilde{\Omega}_j \setminus \Gamma_j$.

The assumptions of Theorem 3.4.1 are satisfied; hence, each v_j has the regularity $(v_j)_t \in L^2(\Omega_j \times (0,T))$, $v_j \in L^2(0,T;W^{2,2}(\Omega_j))$. From the formula

$$\theta = \sum_{j=1}^n v_j$$

it follows that $\theta_t \in L^2(\Omega \times (0,T))$, $\theta \in L^2(0,T;W^{2,2}(\Omega))$. Consequently, we may use Theorem 3.4.2 because we have the regularity of the coefficients needed from the assumptions and obtain $\nabla v_j \in L^2(0,T;L^\infty(\Omega_j))$ for each j. Hence $\nabla\theta \in L^2(0,T;L^\infty(\Omega))$, which is what we wanted to prove. ∎

3.6 Proof of continuous data dependence

Let the hypotheses of Theorem 3.2.5 hold. In terms of (θ_i, u_i), equations (3.2.1)–(3.2.2) have the form, for a.e. $t \in (0,T)$:

$$\int_\Omega ((\theta_i)_t\,\varphi + \nabla\theta_i \cdot \nabla\varphi - R(\theta_i, u_i)\,\varphi)\,dx + \int_{\partial\Omega} h(x, \theta_i, \theta_{\Gamma i}(x,t))\,\varphi\,dS = 0 \quad (3.6.1)$$

$$\int_\Omega (G(\theta_i, u_i)_t\,\psi + \nabla u_i \cdot \nabla\psi - H(\theta_i, u_i)\nabla\theta_i \cdot \nabla\psi)\,dx + \int_{\partial\Omega} b_i(x,t)\psi\,dS = 0 \quad (3.6.2)$$

for every test functions $\varphi, \psi \in V$, where G, H, and R are defined by the identities

$$F(\theta, G(\theta, u)) = u, \quad H(\theta, u) = \frac{\partial F}{\partial \theta}(\theta, G(\theta, u)), \quad R(\theta, u) = r(\theta, G(\theta, u)). \qquad (3.6.3)$$

Hypothesis 3.2.2 implies that G, H, R are Lipschitz-continuous in both variables, $1/d_1 \leq \partial_u G \leq 1/d_0$.

3.6. PROOF OF CONTINUOUS DATA DEPENDENCE

Set
$$U_i(x,t) = \int_0^t u_i(x,\tau)\,\mathrm{d}\tau, \quad u_i^0 = F(\theta_i^0, c_i^0), \quad \bar{U} = U_1 - U_2.$$

We consider the difference of the equations (3.6.1) for $i = 1$ and $i = 2$, tested by $\varphi = \bar{\theta}$, integrate the difference of the equations (3.6.2) for $i = 1$ and $i = 2$ from 0 to t, and test by $\psi = \bar{U}_t$.

We denote by C any constant independent of the solutions, and by ε a small parameter, which will be suitably chosen. Since θ_i and $\theta_{\Gamma i}$ are uniformly bounded, we may assume that h is Lipschitz-continuous in θ and θ_Γ.

74 CHAPTER 3. ANALYSIS OF A RELATED QUASILINEAR PARABOLIC SYSTEM

Hence, using (3.4.8) for an appropriate ε, we obtain

$$\int_\Omega \left(\bar{\theta}_t \bar{\theta} + |\nabla \bar{\theta}|^2\right) dx \leq \int_\Omega (R(\theta_1, u_1) - R(\theta_2, u_2))\bar{\theta}\, dx + C\int_{\partial\Omega} (|\bar{\theta}_\Gamma| + |\bar{\theta}|)\bar{\theta}\, dS$$
$$\leq C\int_\Omega (|\bar{\theta}| + |\bar{U}_t|)|\bar{\theta}|\, dx + C\int_{\partial\Omega} |\bar{\theta}_\Gamma|^2\, dS, \qquad (3.6.4)$$

and

$$\int_\Omega (G(\theta_1, U_{1t}) - G(\theta_2, U_{2t}))\, \bar{U}_t(x,t)\, dx + \frac{1}{2}\frac{d}{dt}\int_\Omega |\nabla \bar{U}|^2(x,t)\, dx \qquad (3.6.5)$$
$$= \int_\Omega \left(\int_0^t \left(H(\theta_1, U_{1t})\nabla\theta_1 - H(\theta_2, U_{2t})\nabla\theta_2\right)(x,\tau)\, d\tau\right) \cdot \nabla \bar{U}_t(x,t)\, dx$$
$$\quad - \int_{\partial\Omega} \left(\int_0^t \bar{b}(x,\tau)\, d\tau\right) \bar{U}_t(x,t)\, dS + \int_\Omega (G(\theta_1^0, u_1^0) - G(\theta_2^0, u_2^0))\, \bar{U}_t(x,t)\, dx$$
$$= \frac{d}{dt}\int_\Omega \left(\int_0^t \left(H(\theta_1, U_{1t})\nabla\theta_1 - H(\theta_2, U_{2t})\nabla\theta_2\right)(x,\tau)\, d\tau\right) \cdot \nabla \bar{U}(x,t)\, dx$$
$$\quad - \int_\Omega \left(H(\theta_1, U_{1t})\nabla\theta_1 - H(\theta_2, U_{2t})\nabla\theta_2\right) \cdot \nabla \bar{U}(x,t)\, dx$$
$$\quad - \frac{d}{dt}\int_{\partial\Omega}\left(\int_0^t \bar{b}(x,\tau)\, d\tau\right)\bar{U}(x,t)\, dS + \int_{\partial\Omega} \bar{b}(x,t)\, \bar{U}(x,t)\, dS$$
$$\quad + \int_\Omega (c_1^0 - c_2^0)\, \bar{U}_t(x,t)\, dx\,.$$

Integrating Eq. (3.6.5)–(3.6.4) with respect to t and using the hypotheses on the data, we obtain

$$\frac{1}{2}\int_\Omega |\bar{\theta}|^2(x,t)\, dx + \int_0^t \int_\Omega |\nabla\bar{\theta}(x,\tau)|^2 dx\, d\tau \qquad (3.6.6)$$
$$\leq C\int_0^t\int_\Omega (|\bar{\theta}|+|\bar{U}_t|)|\bar{\theta}|(x,\tau)\, dx\, d\tau + C\int_0^t\int_{\partial\Omega}|\bar{\theta}_\Gamma|^2 dS\, d\tau + \frac{1}{2}\int_\Omega |\bar{\theta}^0|^2(x)\, dx,$$
$$\frac{1}{d_1}\int_0^t\int_\Omega |\bar{U}_t(x,\tau)|^2\, dx\, d\tau + \frac{1}{2}\int_\Omega |\nabla\bar{U}|^2(x,t)\, dx \qquad (3.6.7)$$
$$\leq C\int_\Omega \left(\int_0^t \left((|\bar{\theta}|+|\bar{U}_t|)|\nabla\theta_1| + |\nabla\bar{\theta}|\right)(x,\tau)\, d\tau\right)|\nabla\bar{U}(x,t)|\, dx$$
$$\quad + C\int_0^t\int_\Omega \left((|\bar{\theta}|+|\bar{U}_t|)|\nabla\theta_1|+|\nabla\bar{\theta}|\right)(x,\tau)|\nabla\bar{U}(x,\tau)|\, dx\, d\tau$$
$$\quad + \int_{\partial\Omega}\left(\int_0^t |\bar{b}(x,\tau)|\, d\tau\right)|\bar{U}(x,t)|\, dS + \int_0^t\int_{\partial\Omega}|\bar{b}(x,\tau)||\bar{U}(x,\tau)|\, dS\, d\tau$$
$$\quad + C\int_0^t\int_\Omega |\bar{\theta}(x,\tau)||\bar{U}_t(x,\tau)|\, dx\, d\tau + C\int_\Omega |\bar{c}^0||\bar{U}(x,t)|\, dx\,.$$

3.6. PROOF OF CONTINUOUS DATA DEPENDENCE

Using Hölder's and Young's inequalities, we may rewrite (3.6.6)–(3.6.7) as

$$|\bar{\theta}(t)|_2^2 + \int_0^t |\nabla\bar{\theta}(\tau)|_2^2 \,\mathrm{d}\tau \tag{3.6.8}$$
$$\leq C\left(\alpha(t) + \int_0^t |\bar{\theta}(\tau)|_2^2 \,\mathrm{d}\tau\right) + \varepsilon \int_0^t |\bar{U}_t(\tau)|_2^2 \,\mathrm{d}\tau,$$

$$\int_0^t |\bar{U}_t(\tau)|_2^2 \,\mathrm{d}\tau + |\nabla\bar{U}(t)|_2^2 \tag{3.6.9}$$
$$\leq C \int_\Omega \left(\int_0^t \left((|\bar{\theta}| + |\bar{U}_t|)|\nabla\theta_1| + |\nabla\bar{\theta}|\right)(x,\tau) \,\mathrm{d}\tau\right)^2 \,\mathrm{d}x$$
$$+ C \int_0^t \left((|\bar{\theta}|_2 + |\bar{U}_t|_2)|\nabla\theta_1|_\infty + |\nabla\bar{\theta}|_2\right)(\tau) |\nabla\bar{U}(\tau)|_2 \,\mathrm{d}\tau$$
$$+ C\left(\alpha(t) + \int_0^t |\bar{\theta}(\tau)|_2^2 \,\mathrm{d}\tau\right)$$
$$+ \varepsilon\left(|\bar{U}(t)|_2^2 + \int_0^t \int_{\partial\Omega} |\bar{U}(x,\tau)|^2 \,\mathrm{d}S \,\mathrm{d}\tau + \int_{\partial\Omega} |\bar{U}(x,t)|^2 \,\mathrm{d}S\right),$$

with $\alpha(t)$ defined by (3.2.6). The first two integrals on the right hand side of (3.6.9) will be estimated using Minkowski's inequality

$$\left(\int_\Omega \left(\int_0^t \left((|\bar{\theta}| + |\bar{U}_t|)|\nabla\theta_1| + |\nabla\bar{\theta}|\right)(x,\tau) \,\mathrm{d}\tau\right)^2 \,\mathrm{d}x\right)^{1/2} \tag{3.6.10}$$
$$\leq \int_0^t \left(\int_\Omega \left((|\bar{\theta}| + |\bar{U}_t|)|\nabla\theta_1| + |\nabla\bar{\theta}|\right)^2(x,\tau) \,\mathrm{d}x\right)^{1/2} \,\mathrm{d}\tau$$
$$\leq \int_0^t \left((|\bar{\theta}|_2 + |\bar{U}_t|_2)|\nabla\theta_1|_\infty + |\nabla\bar{\theta}|_2\right)(\tau) \,\mathrm{d}\tau,$$

and Hölder's and Young's inequalities

$$C \int_0^t \left((|\bar{\theta}|_2 + |\bar{U}_t|_2)|\nabla\theta_1|_\infty + |\nabla\bar{\theta}|_2\right)(\tau) |\nabla\bar{U}(\tau)|_2 \,\mathrm{d}\tau \tag{3.6.11}$$
$$\leq C \int_0^t \left(1 + |\nabla\theta_1(\tau)|_\infty^2\right) |\nabla\bar{U}(\tau)|_2^2 \,\mathrm{d}\tau + \varepsilon \int_0^t \left(|\bar{\theta}|_2^2 + |\bar{U}_t|_2^2 + |\nabla\bar{\theta}|_2^2\right)(\tau) \,\mathrm{d}\tau.$$

respectively. Using the inequality $\frac{\mathrm{d}}{\mathrm{d}t}|\bar{U}(t)|_2 \leq |\bar{U}_t(t)|_2$ a.e., we have in (3.6.9)

$$|\bar{U}(t)|_2^2 \leq \left(\int_0^t |\bar{U}_t(\tau)|_2 \,\mathrm{d}\tau\right)^2.$$

For the boundary terms in (3.6.9), we refer to the trace embedding (3.4.8). Using the inequality $\frac{\mathrm{d}}{\mathrm{d}t}|\bar{U}(t)|_2 \leq |\bar{U}_t(t)|_2$ a.e., we have in (3.6.9)

$$|\bar{U}(t)|_2^2 \leq \left(\int_0^t |\bar{U}_t(\tau)|_2 \,\mathrm{d}\tau\right)^2, \quad \int_0^t |\bar{U}(\tau)|_2^2 \,\mathrm{d}\tau \leq \frac{1}{2}\int_0^t (t^2 - \tau^2)|\bar{U}_t(\tau)|_2^2 \,\mathrm{d}\tau.$$

Choosing ε sufficiently small, we thus obtain from (3.6.8)–(3.6.10) the inequality

$$|\bar{\theta}(t)|_2^2 + |\nabla \bar{U}(t)|_2^2 + \int_0^t \left(|\nabla \bar{\theta}|_2^2 + |\bar{U}_t|_2^2\right)(\tau) \, d\tau \qquad (3.6.12)$$
$$\leq C \left(\alpha(t) + \int_0^t (1+|\nabla \theta_1|_\infty)^2 \left(|\bar{\theta}|_2^2 + |\nabla \bar{U}|_2^2\right)(\tau) \, d\tau \right)$$
$$+ C \left(\int_0^t (1+|\nabla \theta_1|_\infty) \left(|\nabla \bar{\theta}|_2^2 + |\bar{U}_t|_2^2\right)^{1/2}(\tau) \, d\tau \right)^2.$$

Inequality (3.6.12) is of the form

$$v(t) + \int_0^t s^2(\tau) \, d\tau \leq C \left(\alpha(t) + \int_0^t \beta^2(\tau) v(\tau) \, d\tau + \left(\int_0^t \beta(\tau) s(\tau) \, d\tau \right)^2 \right), \qquad (3.6.13)$$

with

$$\beta = 1 + |\nabla \theta_1|_\infty \in L^2(0,T), \quad v(t) = |\bar{\theta}(t)|_2^2 + |\nabla \bar{U}(t)|_2^2, \quad s^2(t) = |\nabla \bar{\theta}(t)|_2^2 + |\bar{U}_t(t)|_2^2. \qquad (3.6.14)$$

To estimate $v(t)$ and $s(t)$, we derive below in Lemma 3.6.2 a refined variant of the Gronwall lemma. Recall first the classical Gronwall estimate.

Lemma 3.6.1. *Let $\alpha \in L^\infty(0,T)$ and $\gamma \in L^1(0,T)$ be given nonnegative functions, and let a nonnegative function $v \in L^\infty(0,T)$ satisfy for a. e. $t \in (0,T)$ the inequality*

$$v(t) \leq \alpha(t) + \int_0^t \gamma(\tau) v(\tau) \, d\tau. \qquad (3.6.15)$$

Then for a. e. $t \in (0,T)$ we have

$$v(t) \leq \alpha(t) + \int_0^t \alpha(\tau) \gamma(\tau) e^{\int_\tau^t \gamma(\sigma) \, d\sigma} \, d\tau \leq \sup_{0<\tau<t} \mathrm{ess}\, \alpha(\tau) \, e^{\int_0^t \gamma(\sigma) \, d\sigma}.$$

Sketch of the proof. Multiplying both sides of (3.6.15) by $e^{-\int_0^t \gamma(\sigma) \, d\sigma}$, we obtain the inequality

$$\frac{d}{dt} \left(e^{-\int_0^t \gamma(\sigma) \, d\sigma} \int_0^t \gamma(\tau) v(\tau) \, d\tau \right) \leq e^{-\int_0^t \gamma(\sigma) \, d\sigma} \alpha(t) \gamma(t).$$

The assertion will follow integrating w.r.t. time the inequality above (for a proof see for instance [52]). ∎

Lemma 3.6.1 can be viewed as a result of the fact that the L^∞-norm of the function v is bounded above by its weighted L^1-norm. We now show that an L^p Gronwall estimate still holds if the L^∞-norm on the left-hand side is replaced with an L^p-norm for $p > 1$.

3.6. PROOF OF CONTINUOUS DATA DEPENDENCE

Lemma 3.6.2. *Let $p > 1$ and its conjugate exponent $p' = p/(p-1)$ be fixed, and let $\alpha \in L^\infty(0,T)$, $\gamma_1 \in L^1(0,T)$, and $\gamma_2 \in L^{p'}(0,T)$ be given, γ_2 nonnegative. Let nonnegative functions $v \in L^\infty(0,T)$, $s \in L^p(0,T)$ satisfy for a. e. $t \in (0,T)$ the inequality*

$$v(t) + \int_0^t s^p(\tau)\,\mathrm{d}\tau \le \alpha(t) + \int_0^t \gamma_1(\tau)\,v(\tau)\,\mathrm{d}\tau + \left(\int_0^t \gamma_2(\tau)\,s(\tau)\,\mathrm{d}\tau\right)^p \qquad (3.6.16)$$

Then there exists a constant M such that for a. e. $t \in (0,T)$ we have

$$v(t) + \int_0^t s^p(\tau)\,\mathrm{d}\tau \le M \operatorname*{sup\,ess}_{0<\tau<t} \alpha(\tau). \qquad (3.6.17)$$

Proof. Set $G_2 = \left(\int_0^T \gamma_2^{p'}(\tau)\,\mathrm{d}\tau\right)^{1/p'}$. We fix δ such that for every $t \in [0,T]$ we have

$$\left(\int_{(t-\delta)^+}^t \gamma_2^{p'}(\tau)\,\mathrm{d}\tau\right)^{1/p'} \le \frac{1}{2},$$

and consider first $t \in [0,\delta]$. By Hölder's inequality we have

$$\left(\int_0^t \gamma_2(\tau)\,s(\tau)\,\mathrm{d}\tau\right)^p \le \left(\int_0^t \gamma_2^{p'}(\tau)\,\mathrm{d}\tau\right)^{p-1} \int_0^t s^p(\tau)\,\mathrm{d}\tau \le 2^{-p}\int_0^t s^p(\tau)\,\mathrm{d}\tau.$$

By substitution we obtain instead of (3.6.16), the following estimate:

$$v(t) + \int_0^t s^p(\tau)\,\mathrm{d}\tau \le \alpha(t) + \int_0^t \gamma_1(\tau)\,v(\tau)\,\mathrm{d}\tau + 2^{-p}\int_0^t s^p(\tau)\,\mathrm{d}\tau$$

and it is easy to see how to obtain (3.6.17) from Lemma 3.6.1.

We want to use an induction argument. To this aim, assume that inequality (3.6.17) is proved for $t \subset [0,k\delta]$ with a constant $M = M_k$, and consider $t \in (k\delta,(k+1)\delta]$. We have

$$\begin{aligned}
\left(\int_0^t \gamma_2(\tau)\,s(\tau)\,\mathrm{d}\tau\right)^p &= \left(\int_0^{t-\delta} \gamma_2(\tau)\,s(\tau)\,\mathrm{d}\tau + \int_{t-\delta}^t \gamma_2(\tau)\,s(\tau)\,\mathrm{d}\tau\right)^p \\
&\le 2^{p-1}\left(\left(\int_0^{t-\delta}\gamma_2(\tau)\,s(\tau)\,\mathrm{d}\tau\right)^p + \left(\int_{t-\delta}^t \gamma_2(\tau)\,s(\tau)\,\mathrm{d}\tau\right)^p\right) \\
&\le 2^{p-1}G_2^p \int_0^{t-\delta} s^p(\tau)\,\mathrm{d}\tau + \frac{1}{2}\int_{t-\delta}^t s^p(\tau)\,\mathrm{d}\tau \\
&\le 2^{p-1}G_2^p M_k \sup_{0\le\tau\le k\delta}\alpha(\tau) + \frac{1}{2}\int_0^t s^p(\tau)\,\mathrm{d}\tau.
\end{aligned}$$

The previous estimation yields:

$$v(t) + \int_0^t s^p(\tau)\,\mathrm{d}\tau \leq \alpha(t) + \int_0^t \gamma_1(\tau)\,v(\tau)\,\mathrm{d}\tau + 2^{p-1}\,G_2^p\,M_k \sup_{0\leq\tau\leq k\delta} \alpha(\tau) + \frac{1}{2}\int_0^t s^p(\tau)\,\mathrm{d}\tau$$

Again, we can see how this last estimate, by means of Lemma 3.6.1, yields (3.6.17), for $t \in (k\delta, (k+1)\delta]$. Having proved the assertion for the base case $k = 1$ and the inductive step, we may assert the the proof by induction is concluded. ∎

We are able now to finish the proof of Theorem 3.2.5. Indeed, inequality (3.6.13) has the same form as in Lemma 3.6.2, with $p = 2$, α replaced by $C\alpha$, $\gamma_1 = C\beta^2$, and $\gamma_2 = C^{1/p}\beta$, with v and s given by (3.6.14). The assertion of Theorem 3.2.5 therefore follows from inequality (3.6.13) and Lemma 3.6.2. ∎

Chapter 4

Numerical results

4.1 Introduction

In the present chapter we present some numerical simulations to show how the model developed and analysed in the previous chapters can be applied in concrete situations. In the last decade, the simulation of heat treatments has covered great importance in research as well as in industrial applications, being able to simulate more or less complex three-dimensional geometries. The practice of heat treatment simulations is used as a tool to deliver quantifying information about the characteristic properties of critical component regions regarding previous production steps and the initial state of the material.

Among the papers concerned with gas carburizing, we mention: [9], [24], [39], [44], [45], [46]. Often they are not concerned with the analysis of the underlying model, but they rather use the empirical trial and error method. Currently, there are several possibilities to perform gas carburizing under very different conditions, starting from the choice of the steel, the type of furnace and of gaseous atmosphere. For an overview of the very recent engineering literature dealing with gas carburizing and case hardening, we refer, for instance, to [38], [50] and references therein.

We investigate first the carburization stage and then we pass to consider the effect of an inhomogeneous carbon distribution in the workpiece on the kinetics of the phases under cooling, in comparison with the results obtained in presence of a homogenous carbon distribution.

We conclude bringing together these two main stages of the process, by showing some simulations performed for a two-dimensional configuration, a gear sector, where both carbon diffusion and quenching have been included.

The simulations were performed on the basis of our model:

$$\rho\alpha\frac{\partial\theta}{\partial t} - div\left(k\nabla\theta\right) = \rho L_p p_t + \rho L_m m_t \quad \text{in } \Omega \times (0,T) \quad (4.1.1)$$

$$\frac{\partial c}{\partial t} - div\left((1-p-m)D(\theta,c)\nabla c\right) = 0 \quad \text{in } \Omega \times (0,T) \quad (4.1.2)$$

$$p_t = (1-p-m)g_1(\theta,c) \quad \text{in } \Omega \times (0,T) \quad (4.1.3)$$

$$m_t = [\overline{m}(\theta,c) - m]_+ g_2(\theta,c) \quad \text{in } \Omega \times (0,T) \quad (4.1.4)$$

$$-k\frac{\partial\theta}{\partial\nu} = h(\theta - \theta_\Gamma) \quad \text{on } \partial\Omega \times (0,T) \quad (4.1.5)$$

$$-(1-p-m)D(\theta,c)\frac{\partial c}{\partial\nu} = \beta(c - c_p) \quad \text{on } \partial\Omega \times (0,T) \quad (4.1.6)$$

$$\theta(x,0) = \theta_0 \quad \text{in } \Omega \quad (4.1.7)$$

$$c(x,0) = c_0 \quad \text{in } \Omega \quad (4.1.8)$$

$$p(0) = 0 \quad \text{in } \Omega \quad (4.1.9)$$

$$m(0) = 0 \quad \text{in } \Omega \quad (4.1.10)$$

where

$$\overline{m}(\theta,c) = \min\{m_{KM}(\theta,c), 1-p\},$$
$$m_{KM}(\theta,c) = \left(1 - e^{-c_{km}(c)(M_s(c)-\theta)}\right) H(M_s(c) - \theta). \quad (4.1.11)$$

Denoting with T the end time, we divide the interval $[0,T]$ as it is shown in Figure 4.1, as

$$[0,T] = [t_0, t_1] \cup [t_1, t_2] \cup [t_2, T].$$

We do not take into account in this study the first time interval $[0, t_0]$, corresponding to the heating stage. In any case the heating phase presents no difficulty from the point of view of the mathematics and of the simulations.

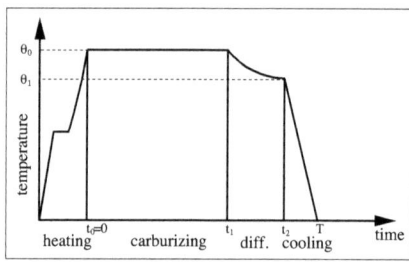

Figure 4.1: Schema of the gas carburizing cycle.

4.1. INTRODUCTION

A general problem consists in determining the material parameters. As we do not have complete experimental data, we proceed as follows: we determine the parameters concerning the phase transitions from the TTT diagrams, for a certain steel, differing by carbon content. By interpolation, we obtain the necessary parameters as function of carbon content. We consider the low-alloyed steel 1320; its TTT diagrams for six different carbon contents (taken from [47], page 17) are shown in Figure 4.2.

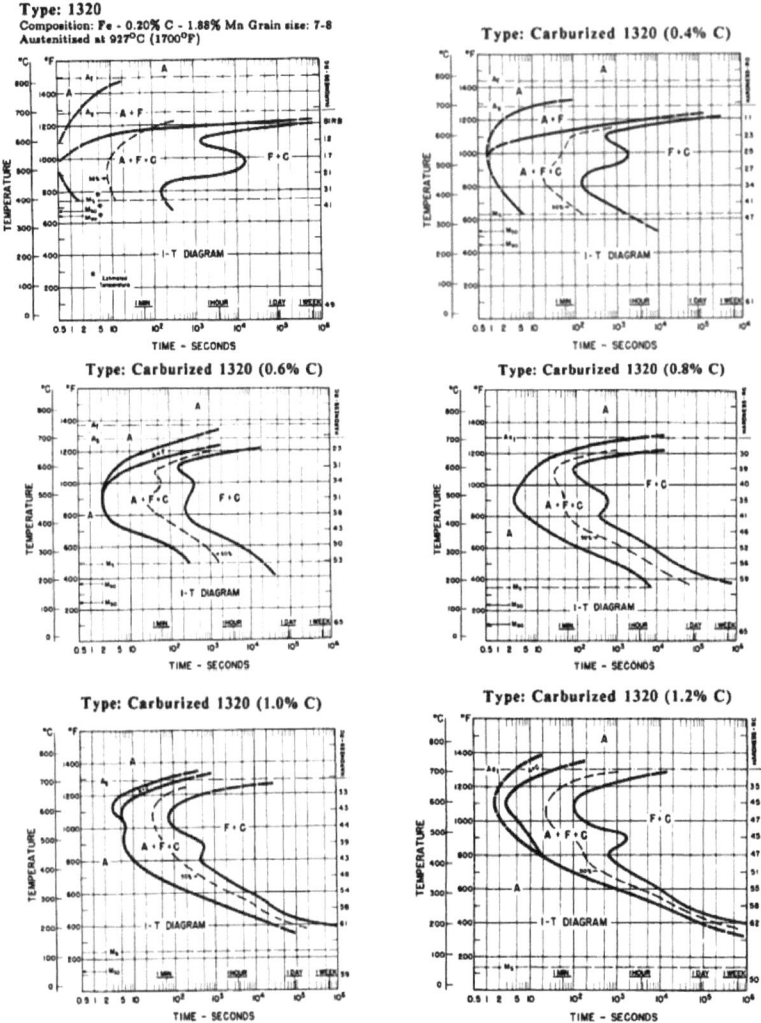

Figure 4.2: TTT diagrams for the steel 1320 for six different carbon contents (from [47]).

From these diagrams we can directly read the values for M_s and M_{90} or M_{50} depending on carbon content, from which we can extract the parameters $M_s(c)$ and $c_{km}(c)$, entering in the formula (4.1.11). They are depicted in Figure 4.3.

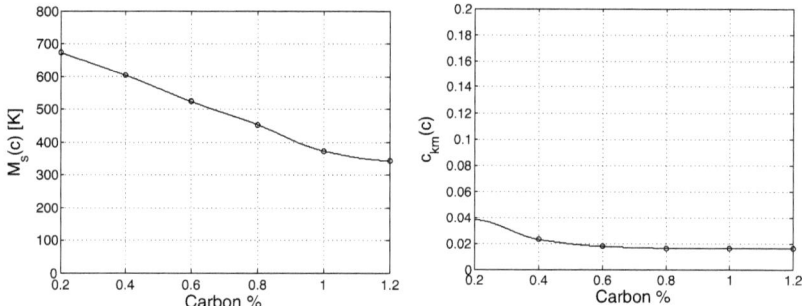

Figure 4.3: Plot of $M_s(c)$ and $c_{km3}(c)$, extracted from the TTT diagrams of Figure 4.2.

We point out that F+C in Figure 4.2 corresponds, in our model, to what we denoted by 'pearlite' (a lamellar structure of ferrite and cementite). A close investigation of the diagrams shows that a small amount of ferrite is formed for carbon content smaller than 0.6%. For carbon content larger that 0.8% there is a prior precipitation of carbides. These two effects are neglected in the following.

The function $g_1(\theta, c)$ in the equation (4.1.3) can be determined with the help of the TTT diagrams of Figure 4.2 arguing as follows. Since no martensite transformation happens above the temperature M_s and no pearlite growth takes place below M_s, in the area $\theta > M_s$, we have $m = 0$.

We take as limits of the integration interval t_s, where pearlite is 1% and t_f, where pearlite is 99%. These two values can be derived from the TTT diagram, for a fixed temperature and for a fixed carbon content. Therefore, for fixed c, it holds:

$$(t_f - t_s)g_1(\theta, c) = \int_{t_s}^{t_f} g_1(\theta, c) dt = \int_{0.01}^{0.99} \frac{dp}{1-p} = [-\log(1-p)]_{0.01}^{0.99} = -\log\left(\frac{0.01}{0.99}\right)$$

$$\Rightarrow g_1(\theta, c) = -\frac{1}{t_f - t_s}\log\left(\frac{0.01}{0.99}\right)$$

The functions so obtained by interpolation between some discrete temperature values, for six different values of c (those occurring in the TTT diagrams), are plotted in Fig-

4.2. CARBURIZATION

ure 4.4 on the left. On the right can be seen a three dimensional plot of the function $g_1(\theta, c)$ by interpolation also in c.

The value for g_2 has been taken after the suggestion of [50] as $1/50$ s^{-1}. A constant value has been found to be sufficient to describe the kinetics of the phase transformations, since temperature and carbon effects are already incorporated in $\bar{m}$.

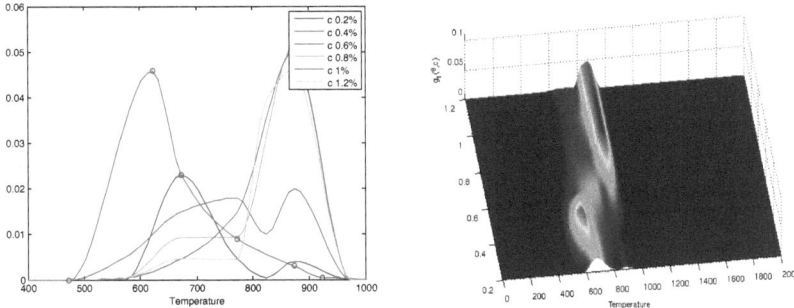

Figure 4.4: g_1 as a function of temperature, for different carbon contents (left). $g_1(\theta, c)$ after interpolation in θ and c.

Further necessary material parameters are listed in Table 4.1.

L_p	L_m	ρ	α	k
77000 J/Kg	82000 J/Kg	7800 kg/m^3	600 J/Kg K	35 W/mK

Table 4.1: Metallurgical parameters for the steel 1320.

The simulations were performed with *Comsol Multiphysics*, a software oriented to the solution of PDEs based on the finite element method. The unit system SI is adopted.

4.2 Carburization

The desired carbon profile is achieved through the carburization and diffusion stages. The two main factors influencing the final case depth of carbon are the *temperature* and the *duration* of carburization. Also significant for the final result is the presence of a diffusion period before quenching. Therefore we focused the simulations on the effect of time and temperature of carburization and on the consecutive diffusion stage. We start considering as sample geometry a one-dimensional domain $\Omega = [0, 0.05]$, assuming the point 0 to be the boundary and the other extreme of the interval the 'inner part' of the workpiece. The initial temperature of carburization, θ_0, is chosen above the austenitization temperature, such that we may assume the workpiece to be homogeneously austenitic, at the beginning.

The values c_0, the initial carbon concentration and c_p, the carbon potential, are both given in weight %. The expression for the diffusion coefficient of carbon in austenite is

$$D(\theta, c) = 0.000047 \exp(-1.6c - (37000 - 6600c)/(1.987\theta)) \text{ m}^2/\text{s}.$$

This formula is taken from Ref. [44] and the value for $\beta = 6 \cdot 10^{-7}$ as well. A plot of $D(\theta, c)$ is depicted in Figure 4.5 (a comparison of different formulas for this coefficient can be found in [41]).

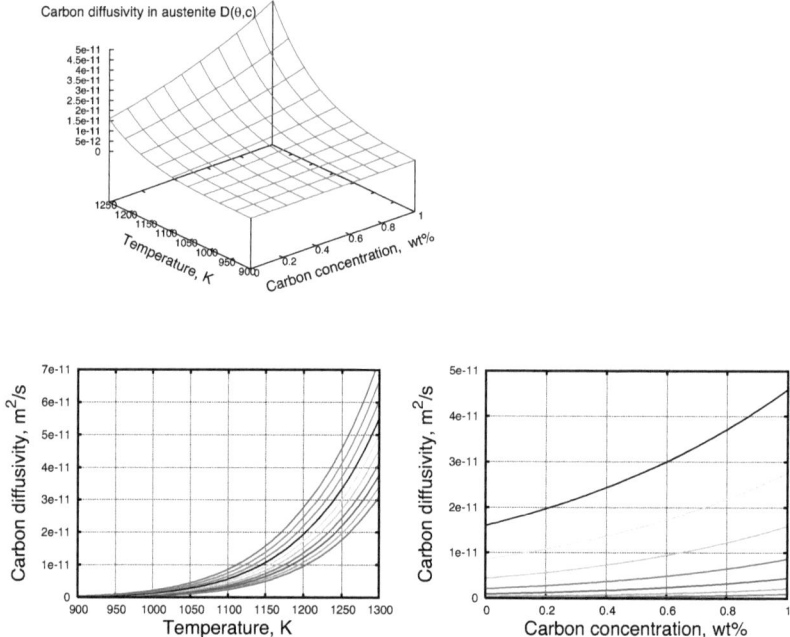

Figure 4.5: Coefficient of carbon diffusion in austenite (upper figure); surface representation as a function of temperature (below on the left), as a function of carbon content (below on the right). According to [44].

4.2. CARBURIZATION

Effect of temperature

Regarding the carburized case, it can vary between a few tenths and several millimitres and must be matched to the function and the size of the component. Its thickness depends strongly on duration and temperature of carburization.

The results, plotted in Figure 4.6 show a good agreement with the ones contained in [44] by C.A. Stickels, which are based on the experiments conducted in [17]. The results obtained in [44] are plotted in Figure 4.7.

The maximum rate at which carbon can be added to steel is limited by the rate of diffusion of carbon in austenite. This diffusion rate increases greatly with increasing temperature; the rate of carbon addition at 1228 K is about 40% greater than at 1144 K. It is mainly for this reason that the most common temperature for gas carburizing is 1200 to 1250 K. For shallow-case carburizing in which the case depth must be kept within a specified narrow range, lower temperature are frequently used, because case depth can be more accurately controlled with the slower carburizing rates obtained with low temperatures.

The indication 'effective case depth' specifies the depth at which the carbon content is 0.4%. Generally, "case depth" is specified as the depth below the surface at which a defined value of some property occurs. A case depth to a hardness of 50 HRC and a case depth to a carbon content of 0.4% are examples of a specifications for an effective case depth.

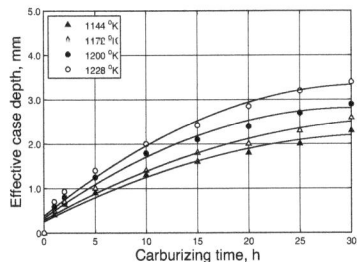

Time, h	1144°K mm	1172°K mm	1200°K mm	1228°K mm
1	0.41	0.49	0.59	0.70
2	0.64	0.70	0.80	0.93
5	0.90	1.00	1.25	1.40
10	1.30	1.40	1.79	2.00
15	1.60	1.80	2.10	2.42
20	1.80	2.00	2.40	2.85
25	2.00	2.30	2.70	3.20
30	2.30	2.59	2.90	3.40

Data derived from the simulations.

Figure 4.6: Plot of effective case depth versus carburizing time at four selected temperatures. Graph based on data in table.

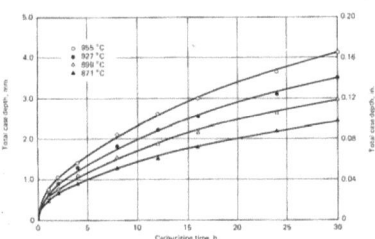

Figure 4.7: Picture taken from [44] depicting the total case depth versus carburizing time at four selected temperatures, based on the experimental data contained in [17].

Effect of time

The depth of case depth achieved versus time is not a linear relationship, but it is more like an inverse proportionality relationship to the fifth power. Referring again to the Table in Figure 4.6, it can be seen that the gain in case depth with increasing time for fixed carburizing temperature is a matter of rapidly diminishing returns. For example, at any of the four temperatures shown, to double the case depth obtained in 1 hour, the time is approximately quintupled.

Effect of the diffusion stage

We examine here the effect of the further carbon diffusion inside the workpiece, during the diffusion period of carbon into the workpiece allowed sometimes before quenching (referring to Figure 4.1, it is the interval $[t_1, t_2]$). The carbon profiles at time t_2 are depicted in Figure 4.8, depending on the length of the interval $[t_1, t_2]$, for different values of t_2.

We recall that the carbon potential assumed in the interval $[0, t_1]$ is $c_p = 1.2$. The green line, which represents the profile of carbon along the interval at time t_1, is a monotonically decreasing function, starting approximately from the value 1.2, as expected. Then this profile falls gradually as t_2 increases in presence of a lower carbon potential c_p and of a lower external temperature θ_{out}; the corresponding profiles after different end times t_2 are indicated with the blue, red and black lines. The rate of decline in this stage is the effect of the diffusion within the steel, the surface reaction rate and the rate at which the atmosphere composition changes. The conditions of the process, specified in Table 4.2, are chosen in compatibility to those used in [45].

4.2. CARBURIZATION

Indeed, the results of our simulations have been compared with those contained in [45], which are shown in Figure 4.9. Here we see the computed carbon gradient in the workpiece at the end of the carburization stage (line 1) and at the end of a diffusion stage of 5000 s. (line 2). The process parameters used by the author of [45], C. A. Stickels, are specified in the same figure. From the comparison between Figure 4.8 and 4.9, we can conclude that the results agree qualitatively and quantitatively.

Time	$t_2 = t_1$	t'_2	t''_2	t'''_2	θ_0	θ_Γ	c_p	β
	8 h	24 h	48 h	72 h	1158 K	800 K	0.8	$2 \cdot 10^{-7}$

Table 4.2: Specified carburizing conditions in the time interval $[t_1, t_2]$ for $t_1 = 8$h.

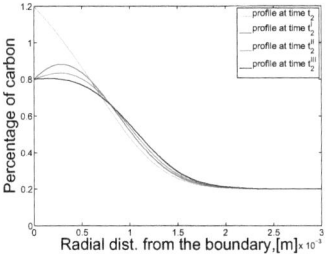 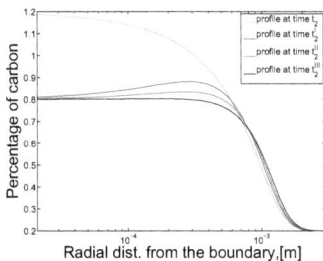

Figure 4.8: Computed carbon gradient at the end of different diffusion times on the lenght of the radius (left) and in an enlargement around the boundary (right).

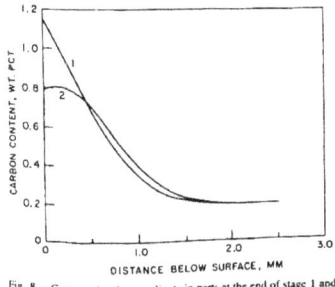

Specified carburizing conditions referred to the figure on the left:

	Stage 1	Stage 2
Temperature, K	2113	1158
Carbon potential c_p	1.15	0.8
β	$2 \cdot 10^{-7}$	$2 \cdot 10^{-7}$
Time, h	4	1.5

Figure 4.9: Picture taken from [45], showing the carbon gradient in a workpiece after carburization (line 1) and additional diffusion (line 2), with the corresponding process parameters in the table on the right.

4.3 Quenching

Now we consider separately the cooling stage of the process, which takes place at a different time scale than the diffusion stage. To this end, we need to specify further process parameters :

h	θ_0	θ_Γ	$T - t_2$
20000 W/m^{2}K	1228 K	300 K	500 s

Table 4.3: Further necessary process parameters.

To understand the effect of an inhomogeneous carbon distribution in the workpiece, we first performed simulations for two homogeneous carbon contents, 0.5% and 0.2%, with $\Omega = [0, 0.05]$. The results are respectively plotted in Figure 4.10 and 4.11. We assumed a heat flux active in x=0 (boundary) while at the other extreme of the interval (inside the body), we assume thermal insulation. The first picture shows the profile of austenite, pearlite and martensite over the interval at the end of quenching; the deviation of the workpiece temperature from the external temperature (T/Tout) is also depicted. The following five pictures represent the phases and temperature evolution in time at five selected points, a, b, c, d, e at increasing distance from the boundary.

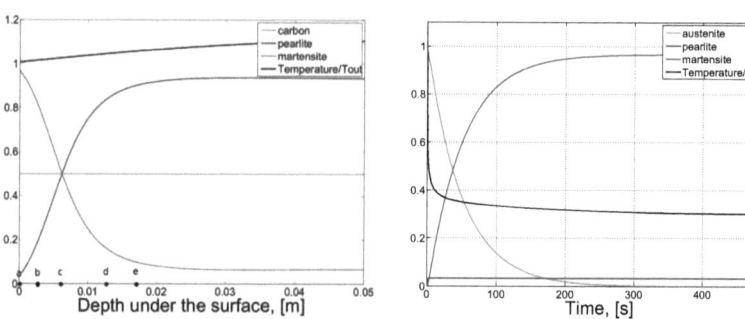

Final phase distributions.   Phases and temperature evolution at point a.

4.3. QUENCHING

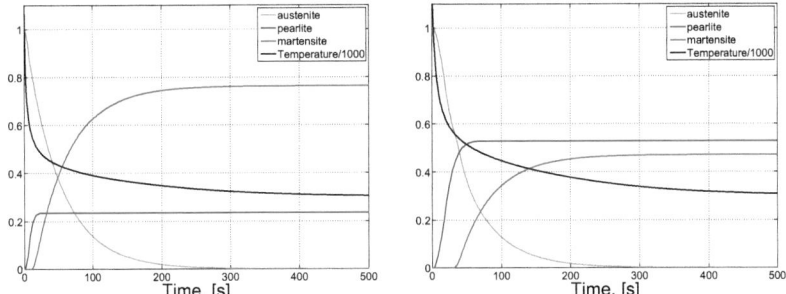

Phases and temperature evolution at point b. Phases and temperature evolution at point c.

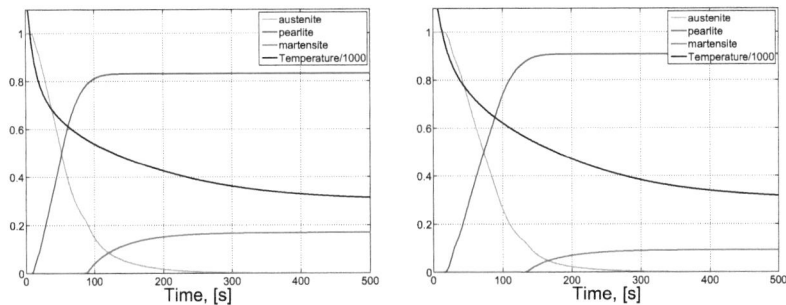

Phases and temperature evolution at point d. Phases and temperature evolution at point e.

Figure 4.10: Final phases distribution for a homogeneous carbon content 0.5% (upper left) and five pictures with the evolution of phases and temperature during quenching at five selected points.

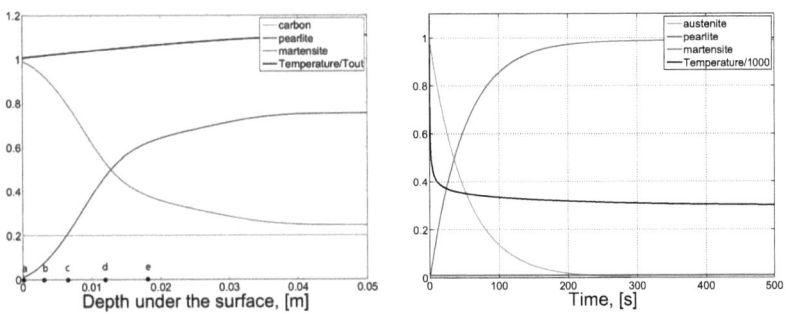

Final phase distributions. Phases and temperature evolution at point a.

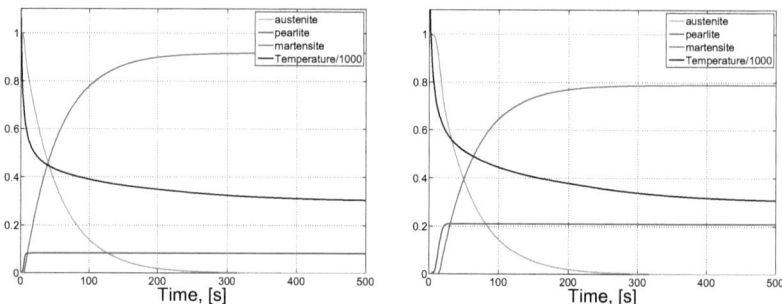

Phases and temperature evolution at point b.   Phases and temperature evolution at point c.

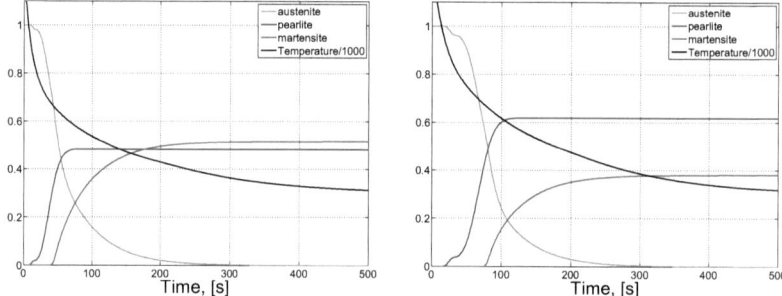

Phases and temperature evolution at point d.   Phases and temperature evolution at point e.

Figure 4.11: Final phases distribution for a homogeneous carbon content 0.2% (upper left) and five pictures with the evolution of phases and temperature during quenching at five selected points.

Through the line depicting T/Tout, it is visible how the cooling rate decreases from the boundary to the interior of the workpiece. Correspondingly, in both cases, we note the usual diversification in the formation of pearlite and martensite: the faster the quenching is, the higher is the amount of marteniste and the lower the one of pearlite formed. We can also see how the phases are monotone function of the spacial coordinate. The kinetic behaviours of the phases in the case of 0.2% carbon and 0.5% carbon are similar, but there are differences in the quantity of phase produced.

Now we pass to the case of a workpiece with inhomogeneous carbon distribution: the carbon profile decreases from 0.5% near the boundary to 0.2% in the middle of the body, which corresponds to the green profile depicted in the first picture of Figure 4.12. As in the previous Figure, we show in Figure 4.12 the resulting final phase

4.3. QUENCHING

distributions over the depth and five plots for the five points selected to observe the time evolution.

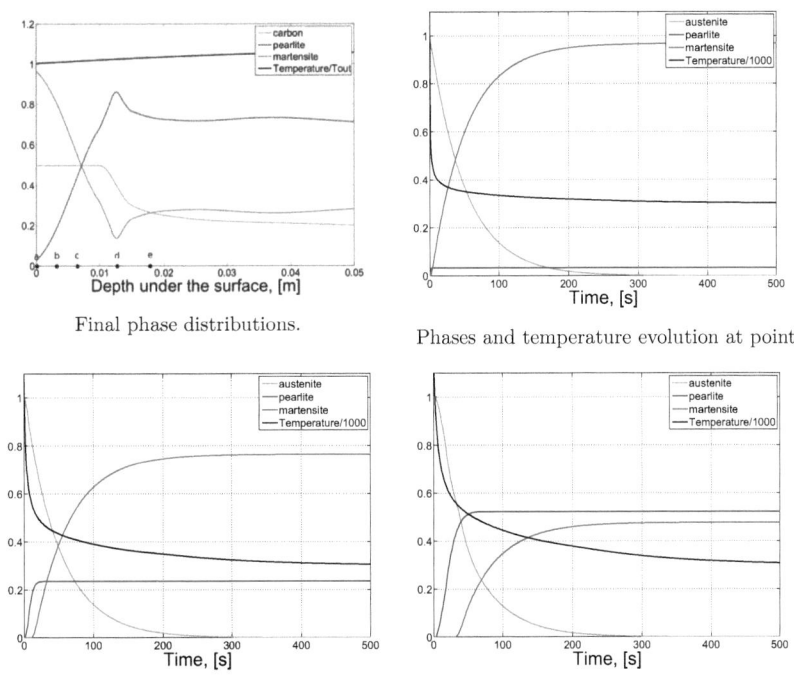

Final phase distributions.

Phases and temperature evolution at point a.

Phases and temperature evolution at point b. Phases and temperature evolution at point c.

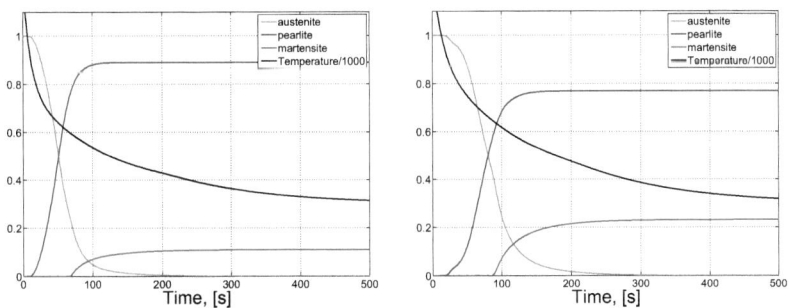

Phases and temperature evolution at point d. Phases and temperature evolution at point e.

Figure 4.12: Final phases distribution for an inhomogeneous carbon profile (upper left) and evolution of phases and temperature at five selected points.

From Figure 4.12 it appears that the final phases distribution is not monotonic with respect to the space, but presents a non monotonicity around the point denoted with d. Looking the pictures in sequence it appears that until the point d the behaviour is similar to the one shown in the case of constant carbon content, i.e., the amount of martensite is high at the boundary, decreasing gradually ; correspondingly the higher amount of pearlite is formed far from the boundary, where the cooling rate is lower. This course is however interrupted passing from point d to point e, where it turns in the opposite direction, as it can be seen from the reduction of the pearlite level.

The phenomenon relies indeed on the differences in carbon content: corresponding to point d is 0.4% and corresponding to e is about 0.27%.

Although a correct interpretation of the phase transformations during cooling would require the continuous cooling diagrams, we still can have some qualitative indications from the TTT diagrams in Figure 4.2. There, one can see that to a decrease in carbon percentage there corresponds a shift to the right of the bulge in the curve denoting the end of the pearlite formation. This results in the fact that more time is needed for the complete formation of pearlite, or, equivalently, in a reduced amount of pearlite producible in a certain interval of time.

In other words, from the diagrams we can infer that when the martensite formation starts, the percentage of pearlite formed at point e is lower than the percentage at the point d. In our pictures we see that, at point d, the maximal percentage of pearlite of almost 90%, whereas at point e this maximum is around 0.75%.

Other carbon configurations between the constant cases 0.2% and 0.5%, could also be wished to suit particular uses; in Figure 4.13 we show three possible profiles and the corresponding final phase distributions.

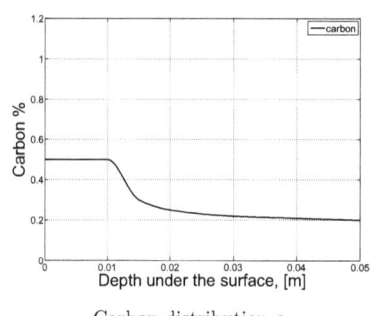

Carbon distribution a.

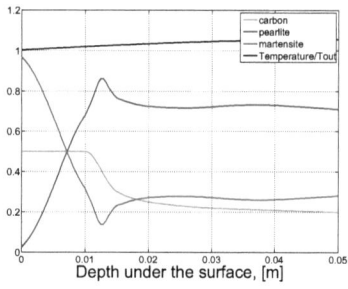

Final distribution, relative to carb. dist. a.

4.4. SIMULATION OF THE COMPLETE PROCESS

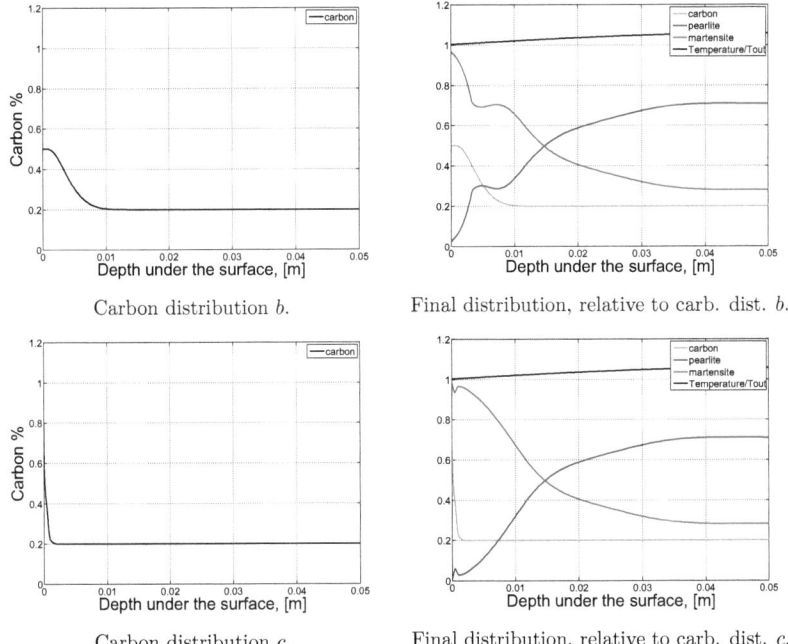

Figure 4.13: Three different carbon profiles with relatives final phase distributions.

4.4 Simulation of the complete process

In conclusion of the chapter, we consider a more complex geometry, namely a gear, which is a mechanical component often subject to case hardening. In this case we performed the simulations of the complete process. For reason of symmetry, it is enough to consider a section of the gear with half a tooth (see Figure 4.14 and 4.15). We employ the same set of parameters as for the one dimensional simulations (T_0=1228 K, h=20000 W/m²K) on the domain Ω is visible in Figure 4.15, of length 14 cm and we suppose that the carbon and the heat fluxes are active on the side of Ω representing the tooth, where the mesh is refined in order to capture what happens in the very thin external boundary layer. We assume $t_1 = t_2$ =10000s and T=10500 s. With these simulations we focus on the effect of the carbon potential, which is a determinant parameter during carburization and, together with the final martensite distribution, determines the final hardness at room temperature.

Four different carbon potentials c_p relevant in the practice (0.2%, 0.4%, 0.6%, 0.8%) were employed. Experimentally it can be seen that the maximal hardenability is achieved indeed for carbon content around 0.8% and that higher carbon amounts do not favour hardness, therefore values of c_p higher than 0.8% are not meaningful. The next simulations show the effect of the different carbon potentials through its influence on the final martensite and carbon distribution. The relation between carbon content, martensite amount and hardness is in Figure 4.21.

Figure 4.15 shows the mesh used and the final temperature distribution. Figure 4.16 is a view of the results of the simulations performed with $c_p = 0.2\%$

Figure 4.14: Steel gear (top) and its horizontal section (bottom).

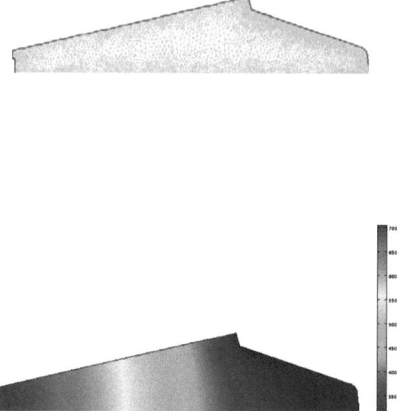

Figure 4.15: Used mesh (top), final temperature distribution (bottom). Colour indicates the temperature value in Kelvin.

4.4. SIMULATION OF THE COMPLETE PROCESS

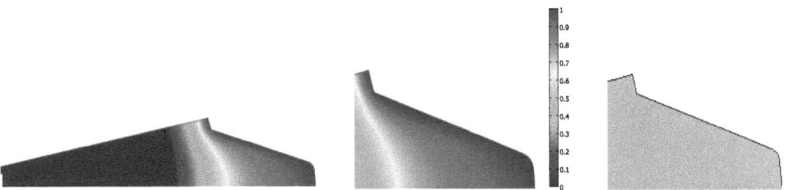

Figure 4.16: Results obtained with $c_p = 0.2\%$: final martensite distribution (left), enlargement (centre), carbon distribution (right) where green colour means 0.2%. The legend refers to the picture in the middle.

Figure 4.16, with the colour green, explains the absence of an effective carbon diffusion, since the steel has 0.2% carbon and 0.2% is also the carbon present in the atmosphere. On the other hand, in Figure 4.17, where the final martensite and carbon distributions are plotted in 0.4%, 0.6%, 0.8% c_p sequence, the carbon distribution, plotted in the right column, is inhomogeneous. The diffusion occurs only in a very thin boundary layer, around 0.5 mm thick and no difference is recognizable for different carbon potentials, whereas slight but visible changes in the marteniste distribution can be seen near the boundary.

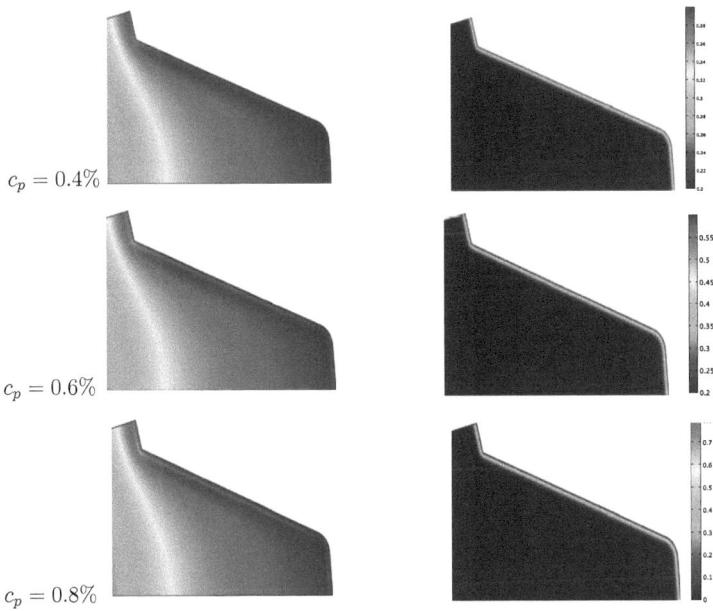

$c_p = 0.4\%$

$c_p = 0.6\%$

$c_p = 0.8\%$

96   CHAPTER 4. NUMERICAL RESULTS

Figure 4.17: Pictures of the final martensite (left column) and carbon (right column) distribution in an enlargement around the carburized layer, starting from the top with carbon potential c_p=0.2% to the bottom with c_p=0.8%.

To enlighten the differences present in the pictures of Figure 4.17, we select now two segments (see Figure 4.18) along which we plot the martensite and carbon percentage, in Figure 4.19.

Figure 4.18: Segments a and b depicted on the total view of the final martensite distribution (obtained with c_p=0.2).

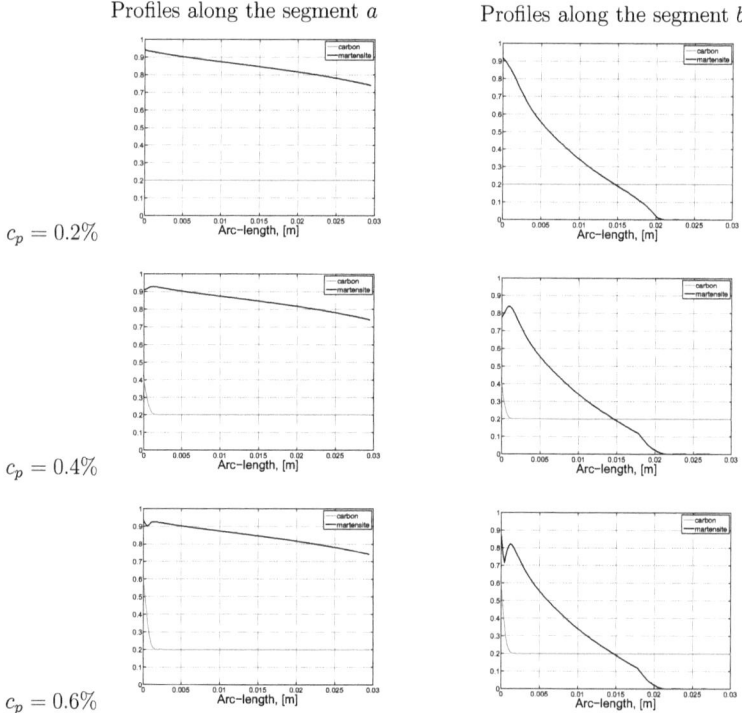

4.4. SIMULATION OF THE COMPLETE PROCESS

$c_p = 0.8\%$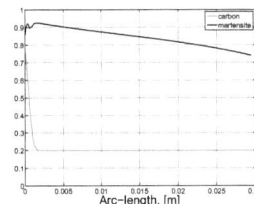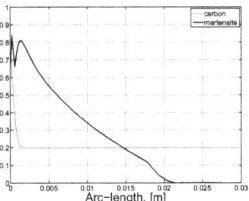

Figure 4.19: Plot of martensite and carbon percentages along the segments a (left) and b (right).

Both segments are 2cm long. Along the segment a, the martensite fraction is 0.9% on the boundary and less lower at the depth of 2 cm; the non monotonic profile is explicable in the same way as in the one dimensional case. Along the segment b the martensite fraction decays more rapidly. This difference between the case a and b reflects the different curvatures in the workpiece.

We conclude with a short consideration about the final hardness. As it can be seen in Figure 4.20, it is possible to achieve different hardening depths, depending on the requests: the grey zone indicates the hardened part. The hardness measurements are obtained experimentally (cf. [5] e.g.). Figure 4.21 shows, schematically, the dependence of hardness on the martensite fraction and on the carbon content. It clearly appears that the maximum hardness is attained for 99% martensite and 0.8% carbon percent. For this reason a carbon potential higher than 0.8% is considered in most cases useless. In other words, a high amount of martensite does not provide high hardness if the carbon level is low. The resulting hardness depth in our two dimensional simulations, in view of the diagram in Figure 4.21, is qualitatively as the one in the right in Figure 4.20, where the entire gear tooth has been hardened.

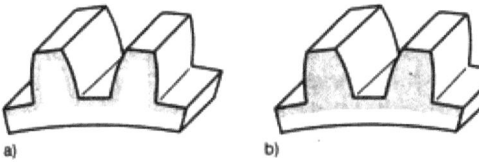

Figure 4.20: Two different hardening depths.

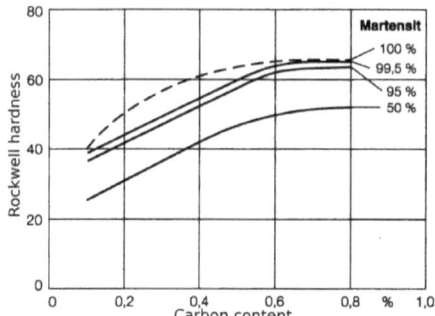

Figure 4.21: Relation between the carbon content (x-axis) and the hardenability (y-axis), depending on the martensite amount after hardening (from [5]).

In conclusion, the purpose of the simulation work of this chapter was to show a possible concrete application of the proposed model. In absence of direct experimental confirmations, we based our simulation on realistic data taken from literature and we have performed numerical simulations using a software based on finite-element method first for a one-dimensional geometry and then for a sector of a gear.

The simulations confirmed, on one side, the expected scenarios in the case of constant carbon content and, on the other side, show interesting phenomena due to the presence of a inhomogeneous carbon distribution, that we have investigated with the help of the time-temperature-transformation diagrams for the considered steel. Under this regard, a cooperation with some engineering institute could be very desirable in order to compare our simulations with the experiments.

Chapter 5

Conclusions

We have modelled carbon diffusion, heat conduction and phase transformations in steel and developed a macroscopic mathematical model for a special variant of case hardening, called gas carburizing. In our model steel was considered as a co-existing mixture of phases, which do not diffuse but rather stay in the position where they have grown. The model allow to describe, from a phenomenological point of view, the kinetics of carburization and the subsequent transformation during cooling. It consists of a coupled system of two parabolic equations for carbon diffusion through the austenite and for heat conduction and ordinary differential equations for the evolution of the phase fractions.

The complete model has been analysed, covering all the stages of the process, and it was shown, under suitable assumptions, the existence of a weak solution through a nested fixed point argument. Under stronger assumptions, also the uniqueness of a solution has been proven.

Next we examined a related problem constituted by a quasilinear system of two parabolic equations with nonlinear boundary conditions, since it was reckoned of intrinsic mathematical interest. Also in this case, we examined the question of the existence of a weak solution of the corresponding initial-boundary value problem and we focused in particular on the question of the uniqueness of the solutions, trying to allow weaker assumptions than those necessary for the whole system. In order to do that, the regularity of the solutions has been investigated. A major achievement is the technique developed to prove the necessary regularity in order to accomplish the proof of uniqueness. We concluded the chapter with an embedding theorem for a wide class of anisotropic Sobolev spaces, that we believe to be of its own interest, which contains as special cases some embedding theorems in the classical Sobolev spaces.

Lastly we have demonstrated how to apply the model developed to a concrete situation: for a one dimensional configuration we have performed numerical simulations. Despite its simplicity, this configuration provides a very useful insight in the comprehension of the behaviour of the phases and of carbon. We selected a specific type of steel for which data were available from literature. Still we had to face the question of how to determine the phase transition rates as functions of the carbon concentration. We bypassed this obstacle using appropriately the available time-temperature-transformation diagrams for the steel considered. The peculiarities due to the presence of an in-homogeneous carbon content have been pointed out and investigated. Finally we showed the simulations performed taking as sample work-piece a gear sector and showed that a thorough description of the process of carburization was obtained.

Future research

There are natural developments for future research both in mathematical sense and in the direction of the applications. From the point of view of the possible application of our model, a co-operation with engineering research institutes would be of great advantage, to enlighten and detail the numerical simulation aspect. The comparisons with the experimental results would require indeed more precise data. The difficulty encountered in this sense, during the development of the thesis, is due to the fact that an exhaustive databank for the material behaviour of steel is still the aim of very active research in the engineering field nowadays and it is not a simple question.

From the point of view of modelling and mathematics, there are other relevant questions that could be investigated. In our study, mechanical deformations are neglected and the distortion, in particular the one related to case hardening, can not be described by the proposed model. The modelling of material behaviour of steel is a large field of current research. Because of the high complexity of some material phenomena of steel, as for example deformations, plasticity, transformation induced plasticity, phase transformation, e.g., it is common practice to study these phenomena by experiments in more or less isolated situations in order to find the basic relations and then to include them in a bulk model of material behaviour, as the one we presented.

In fact, the model treated in the present thesis can be extended via a building-block principle, extending it possibly to a more general material behaviour of steel, including also the mechanical effects. In order to capture the distortion effect, it is necessary to add a momentum balance as well as additional constitutive equations to the model.

For modelling of mechanical behaviour of steel including carbon diffusion and phase transformations, we refer to [26], [27] and references therein.

Another interesting research subject, equally important for the mathematical aspects and the applications, is how to optimize the carburizing process. Indeed, controlling gas carburizing still presents many difficulties, since all the different phenomena occurring on the steel surface have a crucial influence and it is quite complicated trying to drive the whole process to a desired target. Nonetheless, optimization techniques are strongly requested in order to meet quality requirements, by shortening cycle time, enhancing furnace capacity and thereby reducing the energy consumption. In many installations, certain parameters such as the atmosphere composition and the time of exposure are fixed, while the external temperature and the carbon potential are varied to achieve different targets. An ideal objective would be to formulate an optimal control strategy aimed at obtaining a desired carbon profile and a desired temperature at the end of the process. A first approach would consists therefore in minimizing a suitable cost functional subjected to the state equations (1.4.6)-(1.4.15), with control parameters the carbon potential, previously denoted with c_p and the external temperature, θ_Γ. We can expect that, due to the quasilinearity of system (1.4.6)-(1.4.15), in conjuction with the further requests coming form the specificity of the optimal control problem treatment, the analysis would pose many difficulties. Globally this can be foreseen to result in challenging mathematical problem with an immediate relevant application.

Bibliography

[1] H. Amann. *Linear and quasilinear parabolic problems.* Birkhauser, 1995.

[2] D. Andreucci, A. Fasano, M. Primicerio. On a mathematical model for the crystalization of polymers. *in Proceed. 4th Europ. Conf. Math. in Industry* (Hj. Wacker, W. Zulehner eds.), Teubner, Stuttgart, (1991), 3–16.

[3] S. H. Avner. *Introduction to physical metallurgy.* Second Edition, McGraw-Hill Book Company, 1974.

[4] M. Avrami. Kinetics of phase change. *Chem. Phys.*, (1940), 7–9.

[5] H. J. Bargel, G. Schulze *Werkstoffkunde.* Springer, 2005.

[6] O. V. Besov, V. P. Il'in, S. M. Nikol'skiĭ. *Integral Representations of Functions and Imbedding Theorems.* Scripta Series in Mathematics, Halsted Press (John Wiley & Sons), New York-Toronto, Ont.-London, 1978 (Vol. I), 1979 (Vol. II). Russian version Nauka, Moscow, 1975.

[7] H. E. Boyer. *Practical Heat Treating.* American Society for Metals, 1984.

[8] H. Brezis. *Analyse fonctionelle: Théorie et applications.* Masson, 1983.

[9] R. Chatterjee-Fischer. Überblick über die Möglichkeiten zur Verkürzung der Aufkohlungsdauer. *HTM Härterei-Techn. Mitt.*, 40, 1 (1985), 7–11.

[10] J. W. Christian. *The theory of transformations in metals and alloys.* Part I. Pergamon Press, Oxford, 1985.

[11] A. Coddington, N. Levinson. *Theory of ordinary differential equations.* New York, McGraw-Hill, 1955.

[12] R. Collin, S. Gunnarson, D. Thulin. A mathematical model for predicting carbon concentration profiles of gas-carburized steel. *J. Iron Steel Inst.*, (1972), 785–789.

[13] S. Denis, D. Farias, A. Simon. Mathematical model coupling phase transformations and temperature evolutions in steel. *ISIJ International*, 32 (1992), 316–325.

[14] L. C. Evans, *Partial differential equations*. Graduate Studies in Mathematics, V. 19, American Mathematical Society, 1998.

[15] J.A. Griepentrog. Sobolev-Morrey spaces associated with evolution equations. *Adv. in Diff. Eq.*, 12 (2007), 781–840; Maximal regularity for nonsmooth parabolic problems in Sobolev-Morrey spaces. *Adv. in Diff. Eq.*, 12 (2007), 1031–1078.

[16] E. L. Gyulikhandanov, V. V. Kislenkov, S. P. Provotorov. Computing the concentration profile of carbon during the carburizing of steels in controlled natural-gas atmosphere. Translated from *Metallovedenie i Termicheskaya Orabotka Metallov*, 8 (1981), 9–11.

[17] F. E. Harris. Case depth - an attempt at a practical definition. *Met. Prog.*, 44 (1943), 265–272.

[18] M. Hieber, J. Rehberg. Nonlinear parabolic systems with mixed boundary conditions. *SIAM J. Math. Anal.*, 40 (2008), 292–305.

[19] D. Hömberg. A mathematical model for the phase transitions in euctectoid carbon steel. *IMA J. Appl. Math.*, 54 (1995), 31–57.

[20] D. Hömberg, A mathematical model for induction hardening including mechanical effects. *Nonlinear Anal. Real World Appl.*, 5 (2004), 55–90.

[21] D. Hömberg. Irreversible phase transitions in steel. *Math. Methods Appl. Sci.*, 20 (1997), 59–77.

[22] D. Hömberg, A. Fasano, L. Panizzi. A mathematical model for case hardening of steel. *Math. Models Meth. Appl. Sci.*, (2009), in press.

[23] D. Hömberg, W. Wolff. PID-control of laser surface hardening of steel. *IEEE Trans. Control Syst. Technol.*, 14 (2006), 896–904.

[24] H.P. Hougardy. Die Darstellung des Umwandlungsverhaltens von Stählen in den ZTU-Schaubildern. *HTM Härterei-Techn. Mitt.*, 33, 2 (1978), 63–70.

[25] M. Hunkel, T. Lübben, F. Hoffmann, P. Mayr. Using the jominy end-quench test for validation of thermo-metallurgical model parameters. *J. Phys. IV France*, 120 (2004), 571–579.

[26] T. Inoue, T. Yamaguchi, Z. Wang. Stresses and phase transformations occurring in quenching of carburized steel gear wheel. *Mat. Sci. and Tech.*, 1 (1985), 872–876.

[27] T. Inoue, Z. Wang, K. Miyao. Quenching stress of carburized steel gear wheel, *in Proceed. ICRS2*, (G. Beck, S. Denis, A. Simon eds), Elsevier Appl. Sci. London, New York, (1989), 606–611.

[28] H.W. Knobloch, F. Kappel. *Gewöhnliche Differentialgleichungen*. Teubner, 1974.

[29] D. P. Koistinen, R. E. Marburger. A general equation prescribing the extent of the austenite-martensite transformation in pure iron-carbon alloys and plain carbon steel. *Acta Met.*, 7 (1959), 59–60.

[30] A. Kolmogorov. Statistical theory of crystallization of metals. *Bull. Acad. Sci. USSR Mat. Sci.*, 1(1937), 355–359.

[31] A. Koshelev. *Regularity Problem for Quasilinear Elliptic and Parabolic Systems*. Lecture Notes in Mathematics, Springer-Verlag, Berlin, 1995.

[32] P. Krejčí, L. Panizzi. Regularity and uniqueness in quasilinear parabolic systems. *Applications of mathematics*, submitted.

[33] O.A. Ladyženskaja, V.A. Solonnikov and N.N. Ural'ceva. *Linear and Quasilinear Equations of Parabolic Type*. Amer. Math. Soc. Transl. 23, AMS, Providence, RI, 1968.

[34] J. B. Leblond, J. Devaux. A new kinetic model for anisothermal metallurgical transformations in steel including effect of austenite grain size. *Acta Met.*, 32 (1984), 137–146.

[35] G.M. Lieberman. *Second order parabolic differential equations*. World Scientific Publishing Company, 1996.

[36] J.L. Lions. *Quelques méthodes de résolution des problèmes aux limites non linéaires*. Dunod; Gauthier-Villars, Paris 1969.

[37] J.L. Lions, E. Magenes. *Problémes aux limites non homogénes*, Vol II. Dunod, 1968.

[38] J. Lütjens, V. Heuer, F. König, T. Lübben, V. Schulze, N. Trapp. Computer Aided Simulation of Heat Treatment (C.A.S.H.) Teil 2: Bestimmung von Eingabedaten zur FEM-Simulation des Einsatzhärtens. *HTM Z. Werkst. Waermebeh. Fertigung*, 61, 1 (2006) 10–17.

[39] M. Motoyama, R.E. Ricklefs, J.A. Larson. The effect of carburizing variables on residual stresses in hardened chromium steel, *Automotive Engineering Congress and Exposition* Detroit, Michigan, Feb. 24-28. Paper number 750050, (1975).

[40] J. Nečas. *Les méthodes directes en théorie des équations elliptiques*. Academia, Prague, 1967.

[41] A. Ochsner, J. Gegner, G. Mishuris. Effect of diffusivity as a function of the method of computation of carbon concentration profiles in steel. *Met. Sci. Heat. Treat.*, 46 (2004), 3–4 .

[42] J. F. Rodrigues. A nonlinear parabolic system arising in thermomechanics and in thermomagnetism. *Math. Models Methods Appl. Sci.*, 2 (1992), 271–281.

[43] T. Shilkin. Classical solvability of the coupled system a heat-convergent Poiseuille-type flow. *J. Math. Fluid Mech.* 7 (2005), 72–84.

[44] C.A. Stickels. Gas carburizing, *Heat Treatment, Vol. 7, American Society for Metals*, (1997), 312–324.

[45] C.A. Stickels. Analytical models for the gas carburizing process, *Metall. Trans. B*, 20B (1989).

[46] T. Turpin, J. Dulcy, M. Gantois. Carbon diffusion and phase transformations during gas carburizing of high-alloyed stainless steels: experimental study and theoretical modeling, *Metall. and Mat. Trans. A*, 36A (2005), 2751–2759.

[47] G.F. Vander Voort. *Atlas of time-temperature diagrams for iron and steels*, ASM International, 1991.

[48] Verein Deutscher Eisenhütteleute. *Steel, a Handbook for Material Research and Engineering*, Vol. 1-2, Springer, 1993.

[49] A. Visintin. Mathematical models of solid-solid phase transitions in steel. *IMA J. Appl. Math.*, 30 (1987), 143–157.

[50] M. Wolff, C. Acht, M. Böhm, S. Meier. Modelling of carbon diffusion and ferritic phase transformations in an unalloyed hypoeutectoid steel. *Arch. Mech.*, 59 (2007), 1–33.

[51] http://www.lesman.com *Application Note: Carbon Potential Control.* Lesman Company.

[52] E. Zeidler, *Nonlinear Functional Analysis and Its Applications Vol II*. Springer-Verlag, 1990.

Die VDM Verlagsservicegesellschaft sucht für wissenschaftliche Verlage abgeschlossene und herausragende

Dissertationen, Habilitationen, Diplomarbeiten, Master Theses, Magisterarbeiten usw.

für die kostenlose Publikation als Fachbuch.

Sie verfügen über eine Arbeit, die hohen inhaltlichen und formalen Ansprüchen genügt, und haben Interesse an einer honorarvergüteten Publikation?

Dann senden Sie bitte erste Informationen über sich und Ihre Arbeit per Email an *info@vdm-vsg.de*.

Sie erhalten kurzfristig unser Feedback!

VDM Verlagsservicegesellschaft mbH
Dudweiler Landstr. 99   Telefon +49 681 3720 174
D - 66123 Saarbrücken   Fax   +49 681 3720 1749
www.vdm-vsg.de

Die VDM Verlagsservicegesellschaft mbH vertritt

Printed by Books on Demand GmbH, Norderstedt / Germany